International Conference on Sensors, Materials and Manufacturing (ICSMM 2017)

IOP Conference Series: Materials Science and Engineering
Volume 311

Chiayi, Taiwan
24 – 26 November 2017

ISBN: 978-1-5108-5952-4
ISSN: 1757-8981

Printed from e-media with permission by:

Curran Associates, Inc.
57 Morehouse Lane
Red Hook, NY 12571

CURRAN ASSOCIATES INC.
proceedings.com

Some format issues inherent in the e-media version may also appear in this print version.

Printed by Curran Associates, Inc. (2018)

For permission requests, please contact the Institute of Physics
at the address below.

Institute of Physics
Dirac House, Temple Back
Bristol BS1 6BE UK

Phone: 44 1 17 929 7481
Fax: 44 1 17 920 0979

techtracking@iop.org

Additional copies of this publication are available from:

Curran Associates, Inc.
57 Morehouse Lane
Red Hook, NY 12571 USA
Phone: 845-758-0400
Fax: 845-758-2633
Email: curran@proceedings.com
Web: www.proceedings.com

Table of contents

Volume 311

2017 International Conference on Sensors, Materials and Manufacturing (ICSMM 2017)

24–26 November 2017, Chiayi, Taiwan

Accepted papers received: 1 February 2018
Published online: 8 March 2018

Preface

2017 International Conference on Sensors, Materials and Manufacturing (ICSMM 2017)

Organizing Committee

Peer review statement

Papers

Intelligent Materials and Electronic Technology

Nature inspired capacitive sensor with unique and unclonable characteristic

C B Karuthedath and N Schwesinger.....1

An optimized BP neural network based on genetic algorithm for static decoupling of a six-axis force/torque sensor

Liyue Fu and Aiguo Song.....9

Improvement of thermal radiation characteristic of AC servomotor using Al-CNT composite material

Y Kikuchi, H Wakiwaka and M Yanagihara.....18

Smart reconfigurable parabolic space antenna for variable electromagnetic patterns

Sahil Kalra, Rituparna Datta, B.S. Munjal and Bishakh Bhattacharya.....25

High-Performance AC Power Source by Applying Robust Stability Control Technology for Precision Material Machining

En-Chih Chang.....33

Mechanical Design and Manufacturing Engineering

Kinetic analysis of elastomeric lag damper for helicopter rotors

Yafang Liu, Jidong Wang and Yan Tong.....37

Optimal design of geometrically nonlinear shells of revolution with using the mixed finite element method

L U Stupishin, K E Nikitin and A G Kolesnikov.....47

Scheduling of hybrid types of machines with two-machine flowshop as the first type and a single machine as the second type

Ming-Chih Hsiao and Ling-Huey Su.....52

Quality control process improvement of flexible printed circuit board by FMEA

Siwaporn Krasaephol and Parames Chutima.....58

Transportation of part supply improvement in agricultural machinery assembly plant

Anusit Saysaman and Parames Chutima.....66

Seed defective reduction in automotive Electro-Deposition Coating Process of truck cabin

Aekkalag Sonthilug and Parames Chutima.....74

Material Forming and Material Mechanics

Springback law study and application in incremental bending process

Feifei Zhang, Kai He, Xiaobing Dang and Ruxu Du.....79

Ideal flow theory for the double – shearing model as a basis for metal forming design

S Alexandrov and N T Trung.....88

Singular solutions for the rigid plastic double slip and rotation model under plane strain

S Alexandrov and E Lyamina.....96

Surface Engineering and Coatings

One-step microwave plasma enhanced chemical vapor deposition (MW-PECVD) for transparent superhydrophobic surface

Sukrit Thongrom, Yutthana Tirawanichakul, Nantakan Munsit and Chalongrat Deangngam.....103

Evaluation of the residual stresses in 95wt%Al2O3-5wt% SiC wear protection coating using X-Ray diffraction technique

Adel K. Mahmoud, Zaid S. Hammoudi and M.Sc. Student Samah Rasheed.....111

Anti-corrosion properties of coatings with manganese compounds pigmentation

M Ziganshina and E Nurislamova.....117

Building Materials and Structures

Predicting temperature drop rate of mass concrete during an initial cooling period using genetic programming

Santosh Bhattarai, Yihong Zhou, Chunju Zhao and Huawei Zhou.....123

Biocalcification using Ureolytic Bacteria (UB) for strengthening Interlocking Compressed Earth Blocks (ICEB)

M M Zamer, J M Irwan, N Othman, S K Faisal, L H Anneza, A F Alshalif and T Teddy.....133

Mechanical Design of Metal Dome for Industrial Application

Thomas Jin-Chee Liu, Li-Wei Chen and Nai-Pin Lin.....138

Chemical Engineering

Biochar Preparation from Simulated Municipal Solid Waste Employing Low Temperature Carbonization Process

C Areeprasert, P Leelachaikul, G Jangkobpattana, K Phumprasop and T Kiattiwat.....143

A Pontential Agriculture Waste Material as Coagulant Aid: Cassava Peel

N. Othman, N-S Abd-Rahim, S-N-F Tuan-Besar, S. Mohd-Asharuddin and V Kumar.....148

CO_2 Hydrogenation over Ru/χ-Al_2O_3 Catalyst

Kanyanat Jeenjumras, Sunthon Piticharoenphun and Okorn Mekasuwandumrong.....153

Effect of tween 80 on nanoparticle preparation of modified chitosan for targeted delivery of combination doxorubicin and curcumin analogue

Anita Sukmawati, Wahyu Utami, Ratna Yuliani, Muhammad Da'i and Akhmad Nafarin.....158

IOP Conf. Series: Materials Science and Engineering **311** (2018) 011001 doi:10.1088/1757-899X/311/1/011001

PREFACE

It is our great pleasure to welcome you to 2017 International Conference on Sensors, Materials and Manufacturing (ICSMM 2017) which is held in Nanhua University, Chiayi, Taiwan during November 24-26, 2017. ICSMM 2017 is dedicated to issues related to Sensors, Materials and Manufacturing.

The major goal and feature of the conference is to bring academic scientists, engineers, industry researchers together to exchange and share their experiences and research results, and discuss the practical challenges encountered and the solutions adopted. Professors from Taiwan are invited to deliver keynote speeches regarding latest information in their respective expertise areas. It will be a golden opportunity for the students, researchers and engineers to interact with the experts and specialists to get their advice or consultation on technical matters, sales and marketing strategies.

This proceeding present a selection from papers submitted to the conference from universities, research institutes and industries. All of the papers were subjected to peer-review by conference committee members and international reviewers. The papers selected depended on their quality and their relevancy to the conference. The volume tends to present to the readers the recent advances in the field of Sensor Technology, New Materials and Advanced Materials, Automation and Equipment Manufacturing, Manufacturing and Processing etc.

We would like to thank all the authors who have contributed to this volume and also to the organizing committee, reviewers, speakers, chairpersons, sponsors and all the conference participants for their support to ICSMM 2017.

Prof. Edward Yi Chang
National Chiao Tung University, Taiwan
November 25, 2017

IOP Conf. Series: Materials Science and Engineering **311** (2018) 011002 doi:10.1088/1757-899X/311/1/011002

Organizing Committee

Conference Co-Chairs

Prof. Edward Yi Chang, National Chiao Tung University, Taiwan

Prof. Shen-Ming Chen, National Taipei University of Technology, Taiwan

Prof. Kien Wen Sun, National Chiao Tung University, Hsinchu, Taiwan

Program Chairs

Prof. Ashok Srvastava, Louisiana State University, USA

Organizing Chair

Ms. Yang Dan, Hong Kong Chemical, Biological & Environmental Engineering Society, Hong Kong

Publication Chair

Ms. Tammy Luo, Chemistry and Materials Society, Hong Kong Chemical, Biological & Environmental Engineering Society, Hong Kong (CMS-HKCBEES)

International Technical Committee

Prof. Jiujun Zhang, National Reserach Council of Canada, Vancouver, Canada

Prof. Kosai Raoof, University of Maine, France

Prof. Wojciech Wlodarski, RMIT University, Australia

Prof. Luís António Dias Carlos, University of Aveiro, Portugal

Prof. Pavel Ripka, Czech Technical University, Czech Republic

Prof. Solomon W. Leung, Idaho State University, USA

Prof. Songling Huang, Tsinghua University, China

Prof. Tingkai Zhao, Northwestern Polytechnical University, China

Prof. Alexander Gorin, Swinburne University of Technology Sarawak Campus, Malaysia

Assoc. Prof. Xuemin Chen, Texas Southern University, USA

Assoc. Prof. Marcelo H. Prado da Silva, Military Institute of Engineering-IME, Brazil

Assoc. Prof. Yu-Chung Chang, National Changhua University of Education, Taiwan

Assoc. Prof. Xian Luo, Northwestern Polytechnical University, China

Dr. Qifeng Zhang, North Dakota State University, USA

Dr. Lawrence Kulinsky, University of California, USA

Dr. Yang Xi, Diodes Inc., USA

Dr. Spyridon Diplas, Sintef Oslo, Norway

Dr. Subbiah Alwarappan, CSIR-Central Electrochemical Research Institute Karaikudi, India

Peer review statement

All papers published in this volume of *IOP Conference Series: Materials Science and Engineering* have been peer reviewed through processes administered by the proceedings Editors. Reviews were conducted by expert referees to the professional and scientific standards expected of a proceedings journal published by IOP Publishing.

Nature inspired capacitive sensor with unique and unclonable characteristic

C B Karuthedath[1] and N Schwesinger[2]

1 VTT Technical Research Centre of Finland Ltd., FI
2 Technical University of Munich, GE
E-mail: schwesinger@tum.de

Abstract. Background of this paper is the development of sensors showing a nature like characteristic. The sensor is able to detect excitations on inertia bases and operates capacitive. It consists of a miniaturized interdigitated electrode structure on a printed circuit board, a flexible and conductive membrane of PDMS located in a certain distance above and a certain number of steel balls fixed on top of the membrane. The steel ball distribution is random and the conductivity of the membrane is not homogeneous across the membrane. Due to this double random distribution, no sensor equals the other, although the external geometry is equal. The overall size of the sensor is 4.7mm x 4.7mm x 1.7mm. Tilt, acceleration or magnetic fields are capable of causing forces on the steel balls and therefore relative movements between the membrane and the electrode structures. Due to this movement, capacity changes of the arrangement are measurable. This paper describes besides the fabrication of conductive membranes the preparation of regarding sensors. Process technology makes cloning of the sensors impossible. Although all process steps are suited for mass production, no sensor equals the other. Measurements with these sensors prove that each sensor reacts differently to the same excitation. Calculations of the Intra-Concordance-Coefficient show the similarity of the sensors for equal excitations. On the other hand, the maximum Inter-Concordance-Coefficient reveals the differences of such sensors very clearly. Such a characteristic, i.e. equal reaction to equal excitation and an output of significantly different signals allows considering each sensor as a unique device. The sensors obviously behave like receptors in natural organisms. These unusual properties of uniqueness and impossibility to clone make the sensors very interesting for highly secure identification demands. In combination with a very simple measurement procedure, the sensors are an attractive hardware base for technical security solutions.

1. Introduction

While Si-based sensors for one special application possess of an extremely high equality, application specific receptors in nature are similar, but never the same [1, 2]. This non-equality allows for an easy identification. Biological organisms can localize easily for instance, the point of mechanical excitation on the skin due to the knowledge of the signals submitted by the excited effector. With other words, an effector behaves like a fingerprint. In view of security reasons, a clear identification is extremely desirable to prevent unwanted criminal actions. It is not without reason that body characteristics such as the fingerprint or the iris scan are often used to achieve a secure identification. This clear and reliable identification mechanism in nature brought us to the idea to develop sensors with a regarding behavior. Hair cells for instance are similar but never equal. We decided to develop hair-cell-sensors allowing for the detection of inertia-based forces, like acceleration or tilt. A simple way to read out signal is the measurement of capacity changes. This principle is basis of our development. In a first step, we considered the design of the sensor structure. In many cases capacitive sensors have two electrodes located in an opposite position. One electrode is fixed the other is movable. The capacity changes due to the movement of the mobile electrode. This design requires at least contacts at both electrodes. Often this demand leads to complicated design structures. Therefore, we decided to develop a new design. It consists of an interdigitated electrode structure on a printed circuit board (PCB) [3]. Above these electrodes in a certain distance, a membrane of flexible PDMS (Poly-Di-Methyl-Siloxane) is located. The membrane additionally contains several balls of steel. Applying a Voltage to the electrodes, these conductive balls take a certain electrical potential, i.e. they interact with the electrodes. Thus, the arrangement has a certain capacity. A movement of the membrane towards the electrodes or backwards changes the potential and therefore changes the value of the capacity. Since the membrane as structural element is of flexible PDMS, inertia forces on the steel balls can move the membrane when pressure, acceleration or inclination act on the whole system. Figure 1 shows a sketch of the system as discussed.

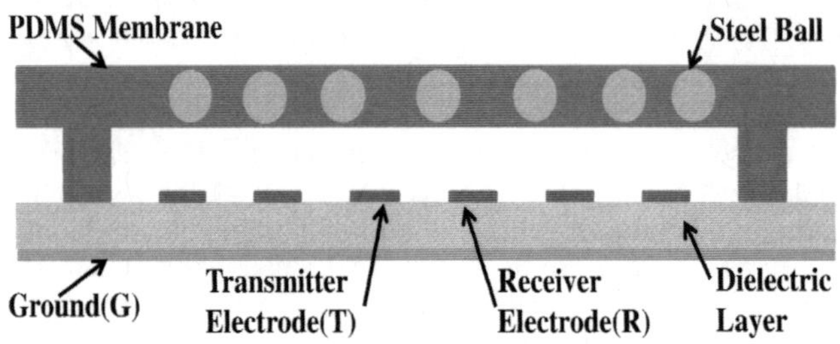

Figure 1: 2D-sketch of the inertia-sensor for measuring mechanical parameters [3]

Transmitter electrode (T) and receiver electrode (R) on the dielectric layer present the interdigitated electrode structure. When steel balls are randomly distributed, the distribution of the mass of structural element is unique. Each steel ball takes a floating potential and acts as a floating electrode. Therefore, the field pattern is also unique for different sensors. Unique field pattern and mass distribution offer a unique capacity output for each sensor. A sophisticated electrode design increases the sensitivity of such a sensor. Tests under pressure, tilt and acceleration show that the sensitivity is comparable with regarding sensors on the market [3]. One main feature of those sensors is the different signal output for equal excitations, a property unwanted in technical solutions. On the other hand, this property makes each sensor unique and comparable with hair cells. Nevertheless, the overall dimensions of such sensors are great (upper mm-range) and not acceptable for technical identification purposes. Therefore, we focused our investigations on miniaturization on one-hand and security demands on the other hand. The phrase "Internet of Things" (IoT) makes the increasing networking of technical systems clear. Unfortunately, this networking has many weak points and allows for unauthorized access to data and devices. Security of systems and devices is therefore a rapidly growing challenge in the modern technical world. A huge number of publications shows the importance of this task. Most solutions have built-in PUF's (Physical Unclonable Function). Many authors focus on electronic PUF's in their work [4-13]. Main disadvantage of all systems is their static design. Often, this is the weak point that makes at least an unauthorized access possible. A PUF-sensor instead, generates a unique response at an external excitation or a challenge [14]. Responses of PUF-sensors are difficult to model and duplicate. Biomimetic describes the utilization of sensors with unique construction and characteristic, i.e. an inherent PUF [12]. Our first development [3] offers all wanted properties in view of security, but unfortunately, it was inacceptable in view of dimensions. Therefore, we focused our work on shrinking the dimensions and improving the security by adding new features to the sensor.

2. Miniaturized sensor

In the first sensor designs, only conductive balls actively contribute to the sensing effect. They cover 30% of the active volume of the membrane. The remaining 70% consist of inactive PDMS. With a dielectric permittivity of about ε_r = 2.75 one can neglect the sensor effect of 70% of the active moveable volume. An increasing number of conductive balls increases their density in the moveable element and leads to a higher sensitivity. However, it reduces the mechanical flexibility of the membrane and at high densities the uniqueness and therefore the performance of the sensor. A promising possibility instead, is the use of conductive PDMS as moveable element. In this case, the complete active volume of the functional element contributes to the sensing process, which helps to achieve a higher sensitivity in combination with a simultaneous reduction of the size. Figure 2. Shows a detailed 2D sketch of the miniaturized sensor.

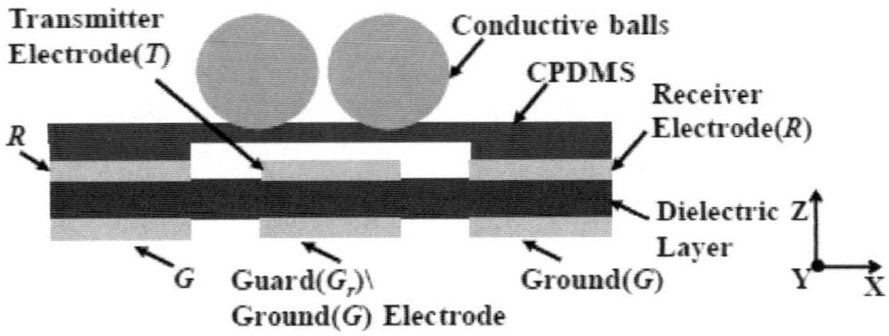

Figure 2: Cross section of a miniaturized sensor

The sensor has a Conductive PDMS (CPDMS) membrane with steel balls fixed on top. The sphere distribution is randomly over the membrane. The dielectric substrate has four electrodes: transmitter (T), receiver (R), ground (G), and ground/guard (Gr). The functional element is in direct contact to the sensor substrate and electrically connected to R. Since the membrane is conductive, it takes the same potential as R. Directly above T, the CPDMS membrane has the active area, i.e. T and the membrane form a parallel plate capacitor. External forces such as pressure, tilt or acceleration deflect the CPDMS membrane up- or downwards. Owing to random ball distribution, the deflection of the CPDMS membrane for the same force is unique for each sensor. This deflection of the CPDMS membrane varies the distance from T, which in turn changes the capacity. G helps to reduce the effect of external interference. The sensor also consists of an additional electrode (Gr) so that the same structure is applicable as a single-electrode or a dual-electrode arrangement.

2.1 Preparation of conductive PDMS

As conductive filling material, we had the choice between CNT (Carbon Nano Tubes) and CB (Carbon Black). For economic reasons we decided to use CB, which is about 20,000 times lower in price than CNT. The mixing procedure was as follows:

I. A required amount of CB was placed in a vessel. Methanol (three times the volume of CB) was added to the CB. Subsequently, the mixture was ultra-sonicated for 15 minutes and then stirred for 1 hour with the help of a magnetic stirrer. During these steps CB was (partially) dissolved in methanol.
II. Directly thereafter, the PDMS base component was added to the CB-Methanol mixture and stirred for 30 minutes. During the process, the mixture began to settle on the bottom of the vessel while the methanol floated above. The excess of methanol was decanted.
III. The regarding Cross-linking Agent (CA) was added to the Methanol-PDMS mixture and stirred for 5 minutes.

For the preparation of CPDMS, we preferred the use of the polar solvent methanol. This solvent has several advantages.

1. Due to a low solubility in PDMS [15], removing of methanol is easy after dispersing the CB filler in CPDMS. Hence, during membrane fabrication, only a very small amount of methanol

remains in the composite, which prevents changes in membrane dimensions due to solvent evaporation during fabrication.
2. Methanol inhibits the cross-linking of PDMS and allows more time for patterning.
3. Methanol is inexpensive and easily available.
4. Considering safety, cost and environmental factors, methanol is listed as a 'recommended' solvent [16]
5. It is possible to reuse an excess of methanol after processing. This makes the fabrication process additionally cost-effective and environmentally friendly.

2.2 Characterization of CPDMS-membranes

A simple two-point resistance measurement was used to characterize the CPDMS-membranes. For that purpose, a simple template was used to measure the resistances The CPDMS membranes were placed on top of measuring electrodes. By means of suitable dimensional markings, the measuring area was fixed, in order to ensure that all the samples had the same surface area. A constant DC voltage source V was applied to the membrane, while the current I through the membrane was measured. The membrane resistance R_M can be estimated with Ohm's law ($R_M = V/I$). Knowing the real dimensions, the specific resistivity of the membrane was finally calculated. Figure 3 shows the Resistivity of the membrane depending on the concentration of the filler (CB) in the membrane. With an increase of the CB-concentration, the resistivity expectedly decreases. The percolation threshold occurs at about 5% of CB-content in the composite.

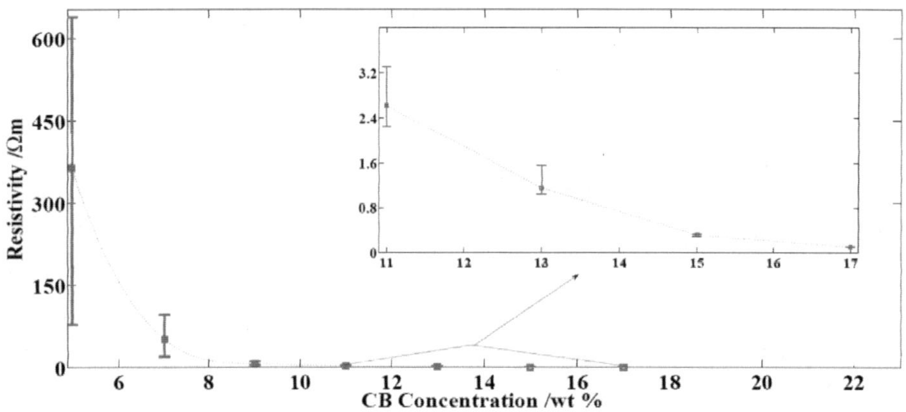

Figure 3: Resistivity of CB-PDMS composites depending on the CB-concentration

Besides the resistivity, the strain behavior of the CPDMS is of interest. For this purpose, we measured the behavior of the membrane depending on an applied mechanical strain. In a tensile test machine, geometrically equal samples were tested. The measurements showed a significant influence of the CB-concentration. The higher the content of CB, the lower is the strain that can be applied. While strain values above about 200% are possible for concentrations up to 5%, we found a breakage of the material at about 140% of strain for 11% of CB and 110% for about 17% of CB. Obviously, the CB reduces the maximum strain of the material and makes the synthetic rubber with an increasing content more brittle. Even the Young's modulus supports this change of properties. At a low content of CB, the Young's modulus is in a range of 0.3MPa. 17% of CB deliver a Young's modulus of about 0.9MPa. This behavior indicates it is not possible to enlarge the amount of CB unlimited. Nevertheless, this behavior says nothing about the homogeneity of the CB concentration. The distribution of CB can be very homogeneous or very inhomogeneous. While homogeneity leads to an equalization of the membranes, inhomogeneity delivers a contribution to uniqueness of each membrane. Further investigations of this task are planned for the near future.

2.3 Preparation of miniaturized sensors

A sketch of the simple and cost-effective fabrication process designed for the miniaturized capacitive sensor shows Figure 4. The process does not require clean rooms or expensive equipment. The CB-PDMS composite was deposited into a master mold and cured to form corresponding CPDMS membranes. Each sensor requires two CPDMS membranes. One of them is the active movable element. The other acts as distance holder between the active element and the PCB. In the active element, up to four conductive balls were randomly distributed. Prior to distribution, the balls were dipped in PDMS, which acted as an adhesive to attach the balls to the CPDMS membrane. A mask was used to limit the ball distribution region within the active area of the functional element. A circular hole was punched through the distance holder-membrane, using a mechanical puncher. During punching, the CPDMS membrane was sandwiched between two protective layers of PDMS. The sensor substrate was fabricated on a double-sided printed circuit board (PCB). Electrode structures were firstly printed on glossy paper and transferred to the PCB afterwards, using a laminating machine. The copper of the PCB was etched to form the electrode structure. The active element, the distance holder and the sensor substrate layer, were aligned manually and bonded together to form the whole sensor structure.

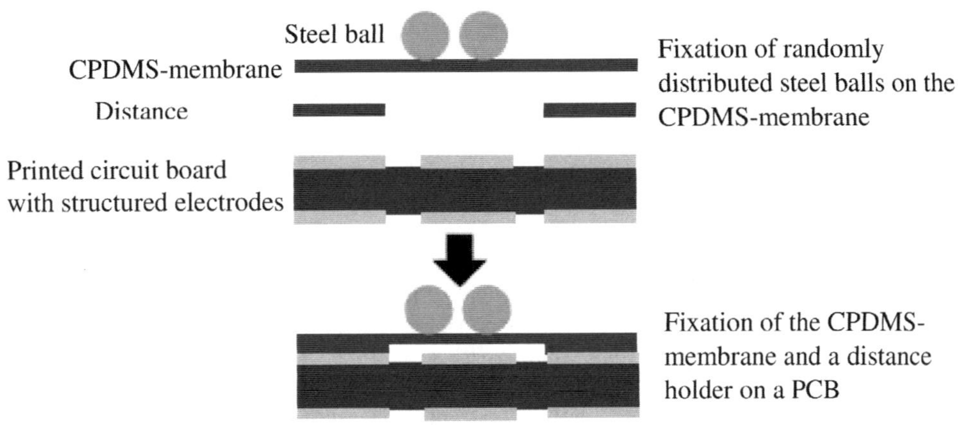

Figure 4: Assembling of different parts of the sensor

3. Experimental setup and results

In order to practically validate the miniaturized sensor design, six sensors (SM_1 – SM_6), each having different ball arrangements, were fabricated. Sensor capacities were measured with an evaluation board. The evaluation board contains an IC AD7746 (Analog Devices), as capacitance-to-digital converter, converting the capacitance into a digital voltage [17]. The IC AD7746 has an accuracy of 4fF, a resolution down to 4aF and can measure changes of the capacity up to ±4pF. These specifications are sufficient to evaluate prototypes of the sensors. A shielded cable connected the sensor electrodes with the evaluation board. Due to that, external interference on sensor output were prevented. A built-in software, available with the evaluation board, recorded the capacities. The performance of the sensors was evaluated with the help of the inclination measurement. During a measurement, the sensor was placed on an inclination platform allowing a tilting from -180° to +180°. The inclination angle was changed in 10° steps. During tilting, the capacitance value of the sensor was recorded. All experiments were carried out under standard laboratory conditions (22°C ± 0.5°C). In order to prove the reproducibility, all measurements were repeated six times with an hour between each. Figure 5 shows repeated (A-F) capacitance values of sample (SM_1) for a full 360°- rotation.

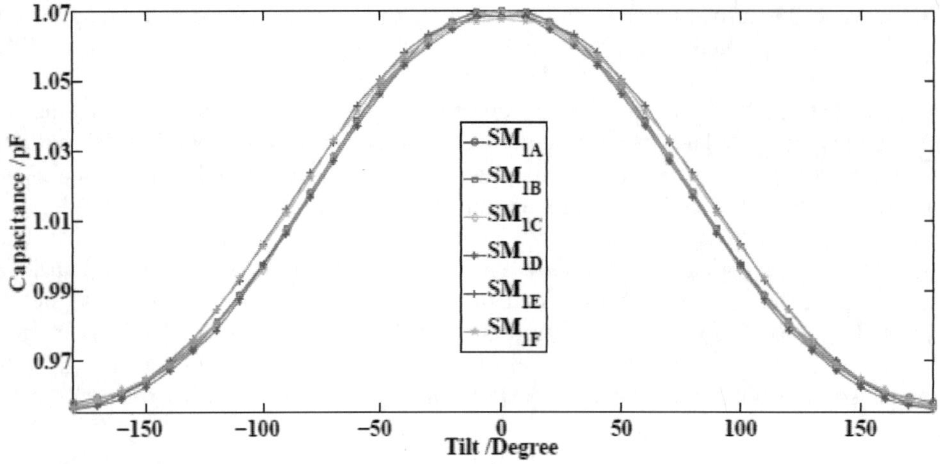

Figure 5: Reproducibility of the measurement values depending on the angle of inclination

3.1 Stability of sensor signals

In order to verify the stability, the sensor output was measured over a longer period. Sensor output at 0° of inclination was recorded for at least 14 minutes (10,000 sample points at a rate of 77ms). The sensor shows an average capacitance of 1.080 pF with a maximum variation uncertainty of 0.003 pF. Humidity, temperature variations, external electromagnetic interferences and noise seem to be responsible to these marginal variations. A further tilt of +90° lead to a new capacitances value. Again, the dates were collected for 14 minutes. A next tilt of additional +90° followed including the long-term measurement. Figure 6 shows the recorded capacitances at three different tilt angles. After changing the tilt angle, it took a certain amount of time (called `settling time') before the output showed stable value. Settling time is considered as the time taken by the sensor output to reach and settle within a tolerance limit (±0.003 pF) of the final value, after tilting. Fig. 6 shows the settling time evaluated from the recorded output. While at +90°, the sensor output stabilized after about 0.5 seconds, it took around 52 s for 180°. This difference in settling times might be influenced by gravity and spring forces on the membrane ordered in different positions. To make sure, the sensor output are stable all records were taken 52 s after tilting.

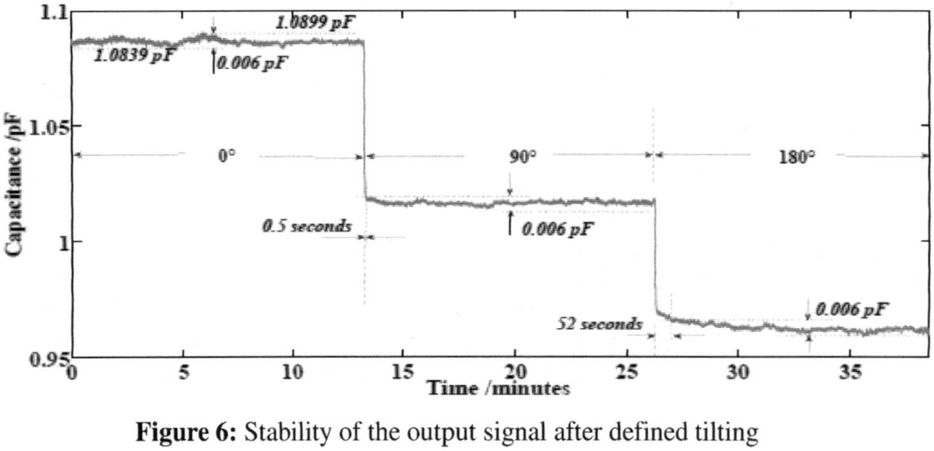

Figure 6: Stability of the output signal after defined tilting

3.2 Uniqueness of the sensors

Since sensors convert certain physical input signals in shape of inclination into electrical signals, all sensors have been tilted from -180° until +180° in steps of 10°. As proof of uniqueness, all sensors had a unique and random distribution of the steel balls. A tilt of each sensor should deliver a signal characterizing the tilt position of the sensor but never equal to all other sensors. In case of a different behavior, each sensor should deliver another output signal at each angle of inclination. Figure 7 shows the corresponding capacity values for six different sensor configurations.

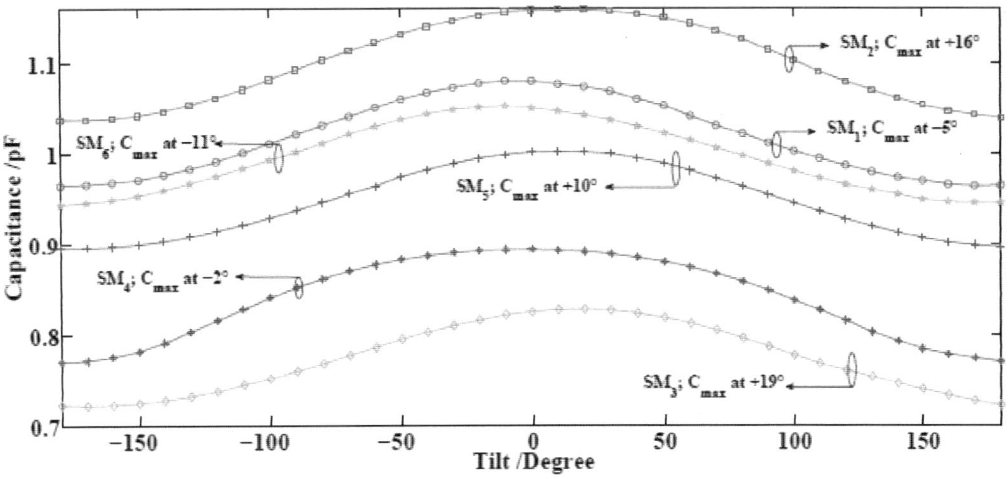

Figure 7: Capacity values of six different sensors depending on the tilt angle

Unexpected but hoped, all sensors show a similar behavior. With the rotation, the values of the capacitance increase and reach their maximum value at approximately 0 °. Then the values decrease again to the initial ones. However, all sensors have different values at the same inclination angles. Therefore, it is very simple to distinguish one sensor from the other. The intra-concordance and the inter-concordance coefficients can be used as a measure of the similarity with simultaneous inequality. In the worst case, the maximum of the inter-concordance coefficient is 0.6500. In the best case, this value would be zero. On the other hand, the similarity of the sensors has in the worst case a minimum intra-concordance-coefficient value of 0.997. The best case would show a value of one. All sensors behave obviously very similar although they show a very high level of uniqueness. These properties allow for a clear identification of each sensor. Sensors like this behave like natural receptors. Since steel balls are magnetically active, it is possible to excite the sensors by magnetic fields, too. Similar to the mechanical excitement, the sensors would react with unique response signals allowing for an exact identification of each sensor. Unique behavior and the impossibility of duplication or cloning makes the devices very attractive in security applications. While security of most identification solutions bases on electronic solutions, the proposed sensor is a real hardware device possessing nature like properties.

4. Summary

A sensor with nature like properties has been developed. They consist of a conductive membrane and randomly distributed steel balls fixed on top of the membrane. The membrane is arranged in a certain distance to an electrode field on a PCB. Forces on the steel balls excite the membrane to move into the direction of the PCB or backwards. Since the membrane in combination with the steel balls acts as a floating electrode, each movement changes the capacity of the arrangement. Mechanical or magnetical forces can initiate such a movement. Although each sensor reacts similar to the excitement, the signal output of each sensor is different. The random steel ball distribution causes this behavior. This unwanted effect in most technical applications is extremely fortunate for security purpose. Similar to receptors in nature, it is possible to identify each single sensor as unique device. The random distribution of steel balls makes it improbable that two sensors deliver the same output signals for an equal excitement.

5. Conclusion

Sensors described above can be considered as identification key. The sensors inherently possess a Physical Unclonable Function - PUF. The integration of such sensors in bank or key cards as well as in other tags could dramatically improve their security against unwanted access. Such hardware solutions are urgently needed especially to ensure the data security of many modern devices in the IoT. To enable integration into corresponding identity cards, our future work is focused on the further miniaturization of these sensors

6. References

[1] M. Burge and W. Burger, "Ear biometrics," in Biometrics. Berlin, Germany: Springer, 1996, pp. 273-285

[2] D. Zhang, Z. Liu, J.-q. Yan, and P.-f. Shi, "Tongue-print: A novel biometrics pattern," International Conference on Biometrics, Crystal City, USA, 2007

[3] B. K. Cyril, and N. Schwesinger, "An Improved Capacitance-Based Intentionally Imperfect Sensor for Measuring Mechanical Parameters", Eight International Conference on Sensing Technology (ICST), 2014, pp. 230–235

[4] Tehranipoor, M., & Wang, C. (2011). Introduction to hardware security and trust. Springer Science & Business Media.

[5] Maiti, A., Casarona, J., McHale, L., & Schaumont, P. (2010, June). A large scale characterization of RO-PUF. In Hardware-Oriented Security and Trust (HOST), 2010 IEEE International Symposium on (pp. 94-99). IEEE.

[6] Bhargava, M., Cakir, C., & Mai, K. (2012). Reliability enhancement of bi-stable PUFs in 65nm bulk CMOS. In 2012 IEEE International Symposium on Hardware-Oriented Security and Trust.

[7] Yin, C. E. D., & Qu, G. (2010, June). LISA: Maximizing RO PUF's secret extraction. In Hardware-Oriented Security and Trust (HOST), 2010 IEEE International Symposium on (pp. 100-105). IEEE.

[8] Yin, C. E., & Qu, G. (2009, July). Temperature-aware cooperative ring oscillator PUF. In Hardware-Oriented Security and Trust, 2009. HOST'09. IEEE International Workshop on (pp. 36-42). IEEE.

[9] Helfmeier, C., Boit, C., Nedospasov, D., & Seifert, J. P. (2013, June). Cloning physically unclonable functions. In Hardware-Oriented Security and Trust (HOST), 2013 IEEE International Symposium on (pp. 1-6). IEEE.

[10] Rose, G. S., McDonald, N., Yan, L. K., & Wysocki, B. (2013, November). A write-time based memristive PUF for hardware security applications. In 2013 IEEE/ACM International Conference on Computer-Aided Design (ICCAD) (pp. 830-833). IEEE.

[11] Kumar, S. S., Guajardo, J., Maes, R., Schrijen, G. J., & Tuyls, P. (2008, June). The butterfly PUF protecting IP on every FPGA. In Hardware-Oriented Security and Trust, 2008. HOST 2008. IEEE International Workshop on (pp. 67-70). IEEE.

[12] Devadas, S., Suh, E., Paral, S., Sowell, R., Ziola, T., & Khandelwal, V. (2008, April). Design and implementation of PUF-based" unclonable" RFID ICs for anti-counterfeiting and security applications. In 2008 IEEE International Conference on RFID (pp. 58-64). IEEE.

[13] Katzenbeisser, S., Kocabaş, Ü., Rožić, V., Sadeghi, A. R., Verbauwhede, I., & Wachsmann, C. (2012, September). PUFs: Myth, fact or busted? A security evaluation of physically unclonable functions (PUFs) cast in silicon. In International Workshop on Cryptographic Hardware and Embedded Systems (pp. 283-301). Springer Berlin Heidelberg.

[14] K. Rosenfeld, G. Efstratios, and K. Ramesh "Sensor physical unclonable functions." IEEE International Symposium on Hardware-Oriented Security and Trust (HOST), 2010, pp. 112-117

[15] J. N. Lee, C. Park, and G. M. Whitesides, \Solvent compatibility of poly (dimethylsiloxane)-based microfluidic devices," Analytical chemistry, vol. 75, no. 23, pp. 6544-6554, 2003

[16] D. Prat, J. Hayler, and A. Wells, \A survey of solvent selection guides," Green Chemistry, vol. 16, no. 10, pp. 4546-4551, 2014

[17] "AD7746 evaluation board: EVAL-AD7746EB datasheet," analog Devices, 2005, USA, Datasheet

IOP Conf. Series: Materials Science and Engineering **311** (2018) 012002 doi:10.1088/1757-899X/311/1/012002

An optimized BP neural network based on genetic algorithm for static decoupling of a six-axis force/torque sensor

Liyue Fu[1] and Aiguo Song[1,2]

1 School of Instrument Science and Engineering, Southeast University, Nan Jing, China
E-mail: a.g.song@seu.edu.cn

Abstract. In order to improve the measurement precision of 6-axis force/torque sensor for robot, BP decoupling algorithm optimized by GA (GA-BP algorithm) is proposed in this paper. The weights and thresholds of a BP neural network with 6-10-6 topology are optimized by GA to develop decouple a six-axis force/torque sensor. By comparison with other traditional decoupling algorithm, calculating the pseudo-inverse matrix of calibration and classical BP algorithm, the decoupling results validate the good decoupling performance of GA-BP algorithm and the coupling errors are reduced.

1. Introduction

For ideal six-axis force/torque sensors, each output value depends only on the force/torque loaded in this direction and not in the other five directions. But actually, for the reason of the mechanical structure of elastic body, the precision of machine operation, strain gauges bonding technology, and measurement methods, etc. almost every force/moment component exerted on sensor impacts on all the output signals, dimensional coupling. Only when dimensional coupling interference is solved, accurate measurement can be achieved.

Dimensional coupling elimination is a key issue for six-axis force/torque sensors. A variety of decoupling algorithms are used to reduce or eliminate the coupling errors. In general, there are two decoupling methods of multi-dimensional force/torque sensors. One method is hardware decoupling, say, the design of 6-axis fore/torque sensors based on mechanism decoupling of the elastic body. Float beam structure [1] and sliding structure [2] are representative mechanical decoupling forms. The other method is software decoupling, decoupling algorithms. Calculating the pseudo-inverse matrix of calibration based on the Least Square Method (LSM) [3] is the most common decoupling algorithm. Due to the relationship between the input and output data of different tunnels is not always linear, Artificial Neural Network is widely used in decoupling algorithms of multi-dimensional force/torque sensors instead of the former method [4]. BP neural network is one of the most common neural networks used for decoupling algorithm [5] [6]. As a result of its local minimum, the decoupling accuracy of BP neural network is barely satisfactory.

With the rapid development of intelligent optimization algorithm, Genetic Algorithm (GA) is widely used in various areas, such as machine learning, pattern identification, control system optimization, and so forth. Aim at overcoming demerits of BP neural network, GA is introduced to optimize the decoupling network in this article.

As a method of seeking the optimum solution by simulating the natural evolution process, GA experts in searching large-scale, non-differentiable, multi-mode space and the gradient information. By doing selection, crossover, mutation and other genetic operations, a new population can be produced. As one of the study hotspots in the intelligent computing, optimized BP neural network based on GA has been widely used in many fields [7][8][9]. Sun wei obtained the index values of the financial security of China's power industry in 2011 according to the tested prediction model based on BP-GA algorithm [10]. Sharon Mano J. Pappu recommended ANN coupled GA to be employed for optimization of fermentative process due to its high accuracy and low prediction error [11]. Jia Weikuan [12] proposed the PLS-GA-RBF algorithm for classification of small sample. Above researchers have achieved good results, but they do not have a systematic description of the BP-GA decoupling model of 6-axis force/torque sensor. Consequently, One decoupling algorithm, optimized BP neural network based on Genetic Algorithm, is proposed in this paper.

2. Six-axis force/torque sensor and static calibration experiment

2.1. Mechanical structure

The mechanical structure of the 6-axis force/torque sensor consists of five parts: bottom cover, pedestal, elastic body, top cover and calibration pillar. The assembled structure of the sensor is shown in Figure 1(a). The performance of the sensor mainly relies on the mechanical structure of the elastic body. As shown in Figure 1(b), the structure of the elastic body is classical cross-beam structure, which is composed of cross beam, compliant beams, a central platform and flange.

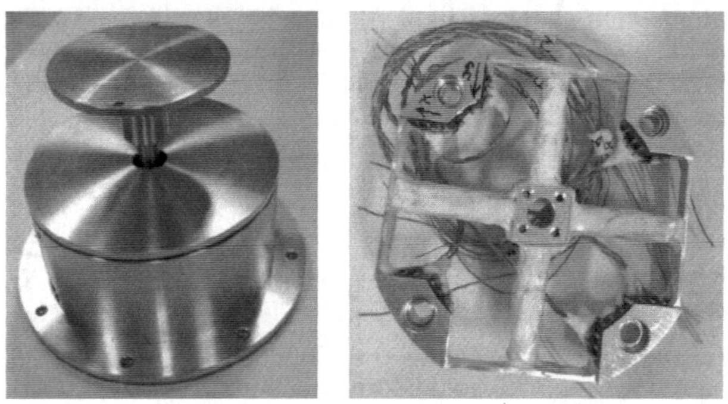

(a) (b)

Figure 1. Prototype of the 6-axis force/torque sensor.

2.2 Calibration experiment

The experiment platform of the calibration tests is illustrated in Figure 2(a). A six-axis force/torque sensor can be fixed to the rotatable indexing plate to guarantee directions of loading forces or torques. Applying a torque in *x*/*y* direction M_x/M_y is shown in Figure 2(b), and Figure 2(c) shows calibration of F_z.

Generally, one force component is applied to the sensor on the condition that other forces/torques are set to zero in the static calibration experiment. However, neural network for decoupling of the six-axis force/torque sensor requires training samples to be more diversified, so as to make training samples provide more information. It is inadequate to carry out calibration test according to the foregoing, and the calibration method adopted in the paper is as follows.

Firstly, the traditional calibration experiments mentioned above are implemented. Secondly, according to the results of the calibration tests, we can roughly get the components with severe coupling interference. Thirdly, do further calibration tests. Exert force/torque on multiple axes with severe coupling interference simultaneously, and record output voltage of each axis. Thus, the calibration data is more diverse, which is very helpful to improve the decoupling accuracy of neural network. The calibration data obtained by applying F_x, in the case of $M_y=0,\ 0.5\ N \cdot m$ respectively, are totally 84 sets of data. Consequently, 504 sets of data were measured in 6 dimensions.

(a) (b) (c)

Figure 2. The calibration experiments setup.

3. Decoupling algorithms and results comparison

Calibration data is fed into the nonlinear decoupling model for decoupling calculation after data preprocessing such as removal of gross errors, systematic error correction, etc. and then the relationships between the input and output signals of the six-axis force/torque sensor are obtained. Two traditional decoupling algorithms are first simply introduced, then GA-BP algorithm is put forward, and their decoupling errors are compared in this section.

3.1. Decoupling algorithm based on calculating the pseudo-inverse matrix of calibration

The output value is proportional to the force or torque exerted in this direction. It can be assumed before analysis that the relationship between the output voltage and the force/torque in other direction is also linear. According to the static calibration experiment data, the input-output relationship can be expressed as shown in equation (1) after ignoring small elements below the order of 10^{-2} in the coefficient matrix.

$$\begin{bmatrix} U_{Fx} \\ U_{Fy} \\ U_{Fz} \\ U_{Mx} \\ U_{My} \\ U_{Mz} \end{bmatrix} = \begin{bmatrix} 1 & 0 & 0 & 0 & 0 & 0 \\ 0 & 1 & 0 & 0 & 0 & 0 \\ 0 & 0 & -1 & 0 & 0 & 0 \\ 0 & 1.732 & -1.34 & 1 & 0 & 0 \\ 0.914 & 0 & 0 & 0 & 1 & 0 \\ -0.239 & -0.137 & 0 & 0 & 0 & 1 \end{bmatrix} \begin{bmatrix} F_x \\ F_y \\ F_z \\ M_x \\ M_y \\ M_z \end{bmatrix} \tag{1}$$

Then, pseudo-inverse matrix of coefficient matrix is worked out, and equation (2) is as follows.

$$\begin{bmatrix} F_x \\ F_y \\ F_z \\ M_x \\ M_y \\ M_z \end{bmatrix} = \begin{bmatrix} 3.2082 & 0 & 0 & 0 & 0 & 0 \\ 0 & 3.0874 & 0 & 0 & 0 & 0 \\ 0 & 0 & -5.0302 & 0 & 0 & 0 \\ 0 & -5.3478 & -6.737 & 4.4643 & 0 & 0 \\ -5.6201 & 0 & 0 & 0 & 4.1859 & 0 \\ 0.7657 & 0.4243 & 0 & 0 & 0 & 2.021 \end{bmatrix} \begin{bmatrix} U_{Fx} \\ U_{Fy} \\ U_{Fz} \\ U_{Mx} \\ U_{My} \\ U_{Mz} \end{bmatrix} \tag{2}$$

Table 1. Decoupling precision of six dimensions.

Coupling error(FS%)	Fx	Fy	Fz	Mx	My	Mz
Type Ⅰ	0.22	0.3	0.36	0.48	0.83	0.56
Type Ⅱ	7.9	8.7	3.8	6.5	7.6	2.8

As can be seen from the equation (2), there is severe dimension coupling relationship between F_x and U_{My}, F_y and U_{Mx}, F_z and U_{Mx}, and so on. Decoupling precision of six dimensions which is calculated from Equation (2) is listed in Table 1. It can be concluded from Table 1that the maximum coupling error of type Ⅰ and type Ⅱ is 0.83% and 8.7%, respectively. The concepts of type Ⅰ and type Ⅱ error are illustrated in appendix section.

3.2. Decoupling algorithm based on BP neural network

The BP neural network is a kind of multilayer feed-forward network based on an error back-propagation algorithm for training. In this paper, architecture of the decoupling BP neural network is 6-10-6, as shown in Figure 3.

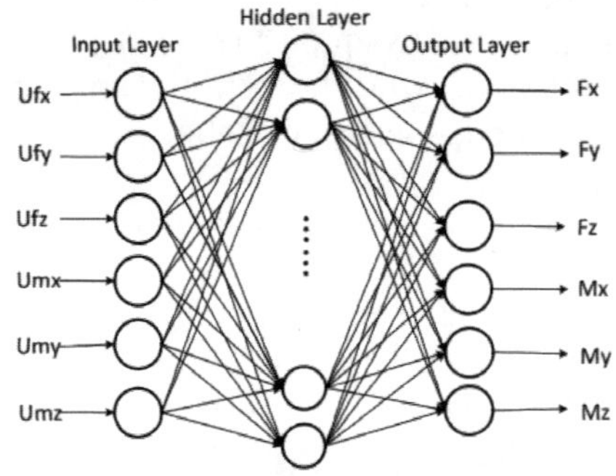

Figure 3. Architecture of the BP-NN model for decoupling.

Table 2. Decoupling precision of BP algorithm in six dimensions.

Coupling error(FS%)	Fx	Fy	Fz	Mx	My	Mz
Type Ⅰ	0.1	0.23	0.34	0.51	0.46	0.12
Type Ⅱ	2.4	2.3	2.5	1.8	2.9	1.4

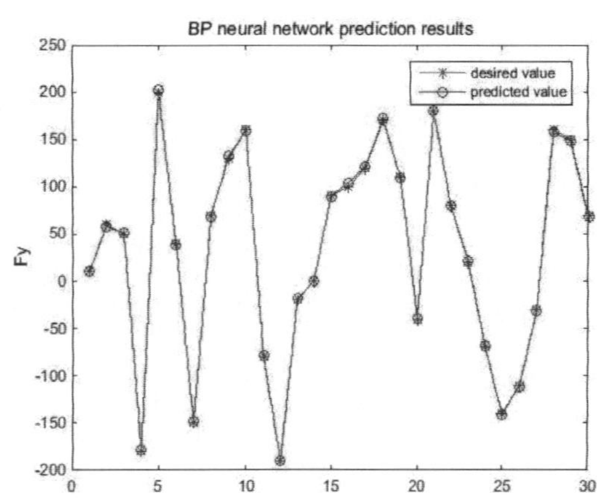

Figure 4. BP neural network prediction results of Fy.

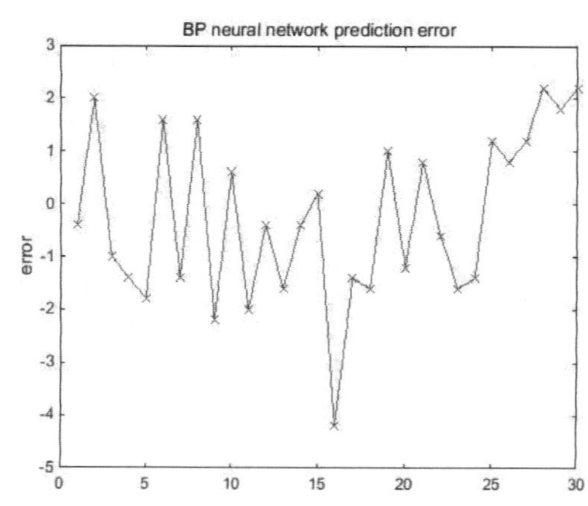

Figure 5. BP neural network prediction error.

A total of 500 sets of data are collected in calibration experiments. 450 of them were used as training samples and the rest were used as test samples. After performance test of the trained neural network, the test results of Fy direction are shown in Figure 4 and 5, and similar in other directions. Decoupling precision of BP algorithm in six dimensions is listed in Table 2. The maximum of coupling error of type Ⅰ and Ⅱ is 0.5% and 2.9%, respectively. The results show that BP algorithm is superior to the decoupling algorithm mentioned in the previous section. Of course, the decoupling process time of this method is longer , reaching 3 seconds or so.

3.3. An optimized BP neural network based on GA

Many people found that the BP algorithm is very sensitive to the initial weights, and the neural network with different generalization performance will be obtained under the only condition that the initial weights are different. Using GA to optimize the initial weights of the neural network can find nearly optimal network connection weights without computing gradient information. Thus, GA is proposed to ameliorate the defect of BP algorithm that is easy to fall into local minimum in this paper.

The basic steps of the optimized BP neural network by GA are as follows:

Step 1 Determine BP neural network topology, and construct this network.

Step 2 Initialize relevant operating parameters of the GA: population size, crossover probability, mutation probability, etc.

Step 3 Create an initial population, and use formula (4) to encode every individual.

Step 4 Calculate the fitness of each individual. Select the inverse of the test error of BP neural network as individual fitness. The flow chart of fitness calculation is as shown in Figure 6.

Step 5 Population evolutions. Population evolutions include selection, crossover and mutation.

Step 6 Determine weather to achieve optimal target. If yes, terminate optimization process of GA; otherwise, turn to step 4.

The best individual in the last population can be used as the approximate optimal solution, i.e. optimal initial weights and biases of BP network. Overall schematic representation of optimized GA-BP neural network is shown in Figure 8.

Table 3. The number of weights and biases in decoupling neural network.

Weights between input layer and hidden layer	Biases of hidden layer	Weights between hidden layer and output layer	Biases of output layer	total
60	10	60	6	136

Table 4. Operating parameters of GA.

Population size	Maximum number of genetic generations	Crossover probability	Mutation probability
40	20	0.6	0.1

3.3.1 Determine encoding mode and encoding length. GA starts from a population with potential solution sets. The population is comprised of a certain number of encoded gene individuals. One of the key issues of constructing genetic algorithm is the encoding method of chromosome. Real number encoding method is adopted in the decoupling model of six-axis force/torque sensor. The complete coding strand of a chromosome is composed of connection weights between input layer and hidden layer, bias of hidden layer, connection weights between hidden layer and output layer, and bias of output layer. The encoding length of individuals is expressed as follows:

$$L = inputnum \times hiddennum + hiddennum + hiddennum \times outputnum + outputnum \quad (3)$$

where L is the coding length of individuals; $inputnum$, $hiddennum$, and $outputnum$ is the number of neurons of input layer, hidden layer and output layer of BP network, respectively.

In view of the architecture of this decoupling neural network, the number of weights and bias is shown in Table 3, and thus the total coding length of chromosome is 136.

3.3.2 Initialize the population. The chromosome of population is encoded as

$$w_{11}^{(1)}w_{12}^{(1)}\cdots w_{1,10}^{(1)}w_{21}^{(1)}w_{22}^{(1)}\cdots w_{2,10}^{(1)}w_{31}^{(1)}\cdots w_{6,10}^{(1)}b_1b_2\cdots b_{10}w_{11}^{(2)}w_{12}^{(2)}\cdots w_{16}^{(2)}w_{21}^{(2)}\cdots w_{10,6}^{(2)}c_1c_2\cdots c_6 \quad (4)$$

where $w_{ij}^{(1)}$, $w_{kl}^{(2)}$ is connection weight from the ith input neuron to the jth hidden neuron and weight from the kth hidden neuron to the lth output neuron, respectively, and b_m, c_n represents threshold of the mth hidden neuron and the nth output neuron, respectively.
Initialize basic parameters of the GA: population size is 40, crossover probability is 0.6, mutation probability is 0.1, maximum number of genetic generations is 20. Operating parameters of GA are initialized as showed in Table 4.

3.3.3 Fitness function. The initial weights and thresholds of BP neural network are obtained by decoding individual. Calculate sum of absolute value of errors between the predicted output and the desired output of the BP network, and then calculate it's reciprocal to be acted as the individual's fitness. The correlation formula is shown as follow:

$$F = \left(\sum_{i=1}^{n} abs(y_i - o_i) \right)^{-1} \quad (5)$$

Here, F is the individual's fitness; n is the number of the output layer neurons of the neural network; y_i and o_i is the expected and predicted output of the ith output neuron, respectively.

3.3.4 Population evolutions. In each generation, individuals are selected according to individual fitness values. With the help of genetic operators, a population representing a new set of solution are produced by crossover and mutation. The roulette wheel is used in selection operation. The basic idea of this method is the probability of individual selection is proportional to the fitness function. The flow chart of population evolutions are shown as Figure 7.

3.3.5 Decoupling results analysis. The fitness evolution curve of the optimized BP decoupling network based on GA is shown as Figure 9. It can be seen from the figure that the prediction error decreases gradually with the evolution, and when the evolution to the 10 generation, the trend of decline is slow. Figure 10 shows the prediction error comparison between BP and GA-BP network. It is easy to see that GA-BP network is better than BP network by comparison of training results.

Table 5. Decoupling precision of optimized BP network by GA in six dimensions.

Coupling error(FS%)	Fx	Fy	Fz	Mx	My	Mz
Type Ⅰ	0.18	0.14	0.21	0.3	0.35	0.1
Type Ⅱ	2.1	1.9	2.2	1.0	2.3	1.3

As can be seen from Table 5, the maximum of coupling error of type Ⅰ and type Ⅱ is 0.35% and 2.3%, respectively. The results show that the proposed decoupling algorithm in this section is superior to BP algorithm in decoupling precision. Althouth this decoupling algorithm has the highest accuracy of the three methods mentioned above, the decoupling time is the longest. It takes about 5 minites to complete a decoupling process and not suitable for online decoupling.

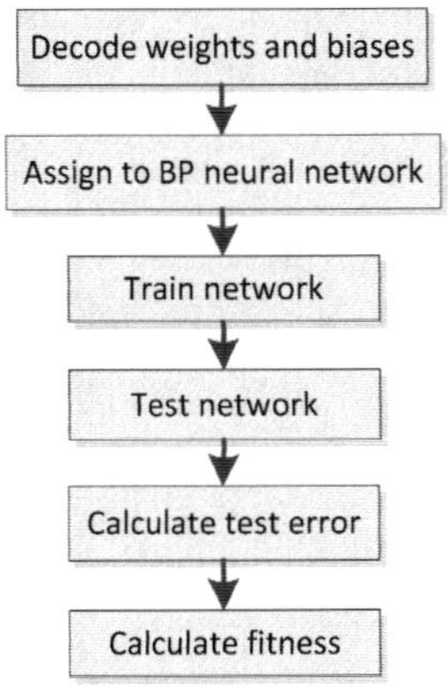

Figure 6. Flow chart of fitness calculation.

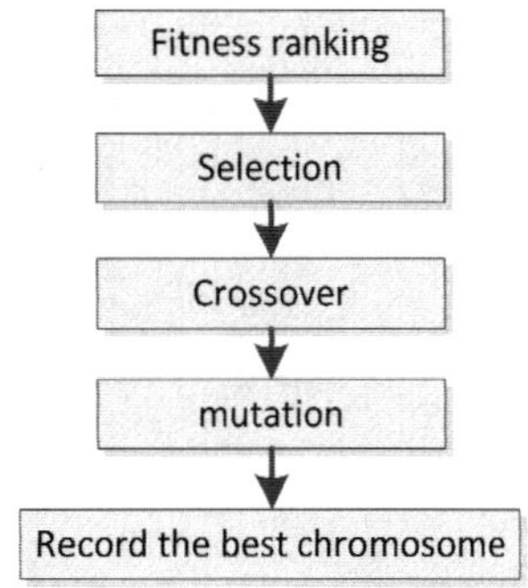

Figure 7. Flow chart of population evolutions.

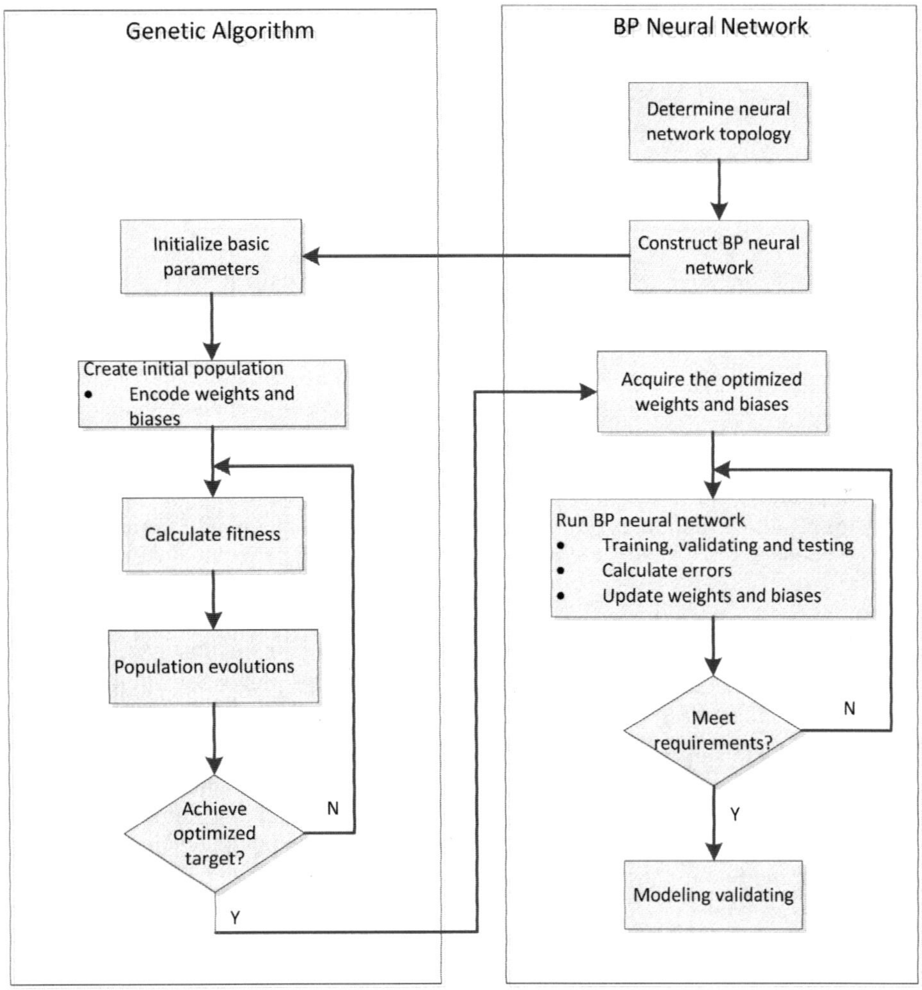

Figure 8. Flow chart of GA-BP algorithm.

IOP Conf. Series: Materials Science and Engineering **311** (2018) 012002 doi:10.1088/1757-899X/311/1/012002

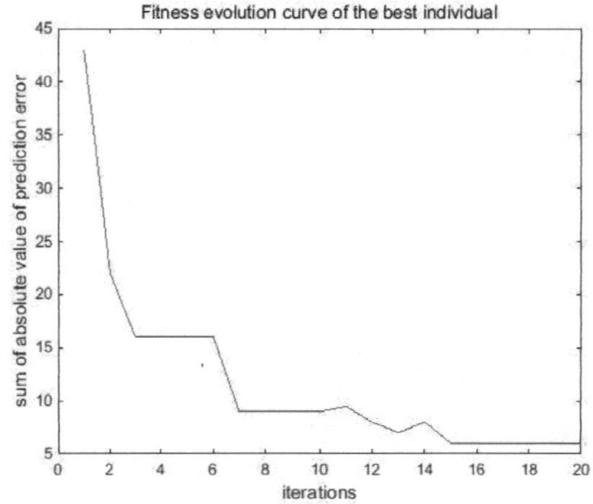

Figure 9. The fitness evolution curve.

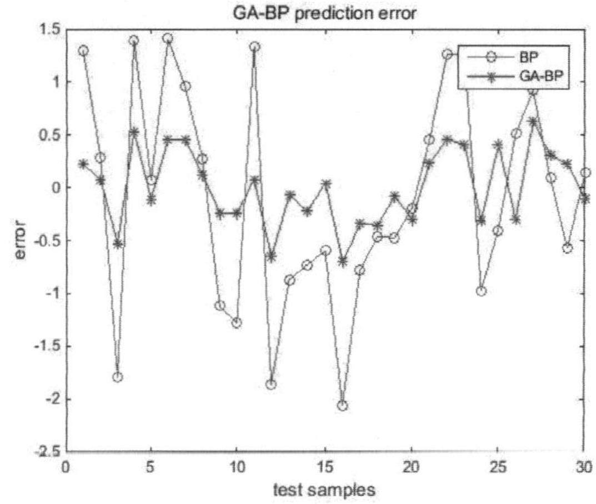

Figure 10. Prediction error comparison between BP and GA-BP network.

4. Conclusion

In the light of the low decoupling accuracy of traditional decoupling algorithms, GA-BP algorithm used for static decoupling of 6-axis force/torque sensor is proposed in this paper. GA is used to optimize the initial weights and biases of the BP neural network. Besides, decoupling algorithm based on calculating the pseudo-inverse matrix of calibration and classical BP algorithm are selected for comparison in order to verify the validity of the proposed algorithm. The maximum coupling error type of Ⅰ is 0.83%, 0.51%, 0.35%, respectively, and the maximum coupling error type of Ⅱ is 8.7%, 2.9%, 2.3%, respectively. The decoupling analysis results indicate that GA-BP decoupling algorithm has the highest accuracy of the three algorithms mentioned in this paper, the decoupling time, however, is the longest. It is appropriate for accurate decoupling offline.

5. Appendices

type Ⅰ: It reflects the degree of deviation between the measured value and the actual applied value of a certain dimension.

type Ⅱ: It reflects the interference size in the direction of *i*, while a force/moment not applied in this direction, whereas applied in other directions.

$$\text{type I}=\left|\frac{e_{i(\max)}}{Y_{i(F.S)}}\right|, \quad i=1,2,\cdots\cdots,6 \tag{A.1}$$

$$\text{type II}=\sqrt{\frac{\sum\left|y_{ij(\max)}\right|^2}{\left|Y_{i(F.S)}\right|^2}}, j=1,2,\cdots\cdots,6, i\neq j \tag{A.2}$$

where $e_{i(\max)}$ is the difference between the measured value and the actual applied value in the direction of i, $Y_{i(F.S)}$ represents the full scale value of the force/torque that can be applied in the direction of i, and $y_{ij(\max)}$ represents the maximum force/torque value measured in the direction of i when the force/torque is applied in the direction of j and not in the other five directions.

6. References

[1] Song Aiguo, Wu Juan, Qin Gang and Huang Weiyi 2007 A novel self-decoupled four degree-of-freedom wrist force/torque sensor *Meas.* **40** 883–91

[2] Wu Bo and Cai Ping 2013 Decoupling analysis of a sliding structure six-axis force/torque sensor *Meas. Sci. Rev.* **13** 187–93

[3] Beyeler F, Muntwyler S *et al* 2009 *IEEE Int. Conf. on Robotics and Automation (Kobe)* pp 520–525

[4] Cao Huibin, Sun Yuxiang, Liu Limin, Feng Yong, Wang Yijun and Ge Yunjian 2011 Coupling analysis of multi-axis force sensor and research of decoupling method *Chinese Journal of Sensors and Actuators* **24** 1136–40

[5] Jiang Li, Liu Hong and Cai Hegao 2004 Nonlinear static decoupling of multi-axis force/torque sensor *Chinese Journal of Scientific Instrument* **25** 284–7

[6] Chen Danfeng, Song Aiguo and Li Ang 2015 Design and calibration of a six-axis force/torque sensor with large measurement range used for the space manipulator. *Proc. Eng.* **99** 1164–70

[7] Yao X 1999 Evolving artificial neural networks. *Proc. IEEE* **87** 1423–47

[8] Venkadesh S and Hoogenboom G 2013 A genetic algorithm to refine input data selection for air temperature prediction using artificial neural networks. *Appl. Soft Comput.* **13** 2253–60

[9] Maarouf M, Sosa A *et al* 2015 The role of artificial neural networks in evolutionary optimization: a review *Computational Methods in Applied Sciences* (Switzerland: Springer) chapter 4 pp 59–76

[10] Sun Wei and Xu Yanfeng 2016 Financial security Evaluation of the electric power industry in china based on a back propagation neural network optimized by genetic algorithm *Energy* **101** 366–79

[11] Sharon Mano J Pappu and Sathyanarayana N G 2017 Artificial neural network and regression coupled genetic algorithm to optimize parameters for enhanced xylitol production by debaryomyces nepalensis in bioreactor *Biochem. Eng. J.* **120** 136–145

[12] Jia Weikuan, Zhao Dean and Ding Ling 2016 An optimized RBF neural network algorithm based on partial least squares and genetic algorithm for classification of small sample. *Appl. Soft Comput.* **48** 373–384

Improvement of thermal radiation characteristic of AC servomotor using Al-CNT composite material

Y Kikuchi[1], H Wakiwaka[2] and M Yanagihara[3]

1 Faculty of Engineering, Shinshu University, 3349-1 Zakoji, Iida, 395-0001, JAPAN
E-mail: yokikuchi@shinshu-u.ac.jp
2 Faculty of Engineering, Shinshu University, 4-17-1 Wakasato, Nagano, 380-8553, JAPAN
3 Faculty of Engineering, Shinshu University, 3349-1 Zakoji, Iida, 395-0001, JAPAN

Abstract. This study deals with a high thermal conductivity material of aluminum-carbon nanotube (CNT) composite with carbon fiber (CF) and the high radiation performance of AC servomotor using a stator made of nanotube composite material. The composite fabrication process was performed by melting a mixture of granular aluminum of less than 200 μm and CNT under conditions of pressed atmosphere at the same time. Two kinds of motors made using aluminum and the composite were evaluated to confirm the effect of thermal conductivity as the motor stator. A test rod of the composite with 14 wt% CF-7 wt% CNT-aluminum indicated the excellent thermal conductivity of 169 W/(mK) in the radial direction and 173 W/(mK) in the lengthwise direction. According to the obtained temperature radiation characteristic of the AC servomotor, the composite stator using CNT decreased the consumption energy to 16% compared to the conventional one. As a result, the highly efficient motor improved the radiation characteristic using the CNT composite stator.

1. Introduction

In recent years, nanotechnology has been hoped to impact industrial products in the future. Especially, nanocarbon material is becoming a central focus of public attention, and it has many structures with regard to form and crystallization, so there is a good possibility of its wide application in the industry as a material with easy manufacturing characteristic. A particularly important and topic material is carbon nanotube (CNT), which is useful with an excellent thermal conductivity of about 1000 W/(mK). CNT is a structure with a six-membered carbon ring, rolled and combined periodically; its perfect form consists of rolled multilayers that form a concentric circle structure. Its outside diameter is dependent on the number of layers and fabrication methods using physical vapor growth. The structure is classified as one layer of single-walled CNT (SWCNT), two layers of double-walled CNT (DWCNT), and a multi-walled CNT (MWCNT) [1]. Research and development have started from vapor growth carbon fiber (VGCF) 100 years ago. In the 1800s, it was found that VGCF with slender fiber form was made by thermal decomposition method from CH_4 or C_2H_2. In the 1990s, it was first made by Hyperion Catalysis Internal, Inc. and Showa Denko K.K. [2] using the industrial process. There are three kinds [3] of manufacturing processes, which are physical vapor deposition (PVD), chemical vapor deposition (PVD), and organic precursor route method. The fabrication methods are similar to that of the CF.

The results of CNT are applicable to the body constructive material of an airplane [4] as a composite with resin mixed CNT and a Li-ion electric battery [5] to store the energy.

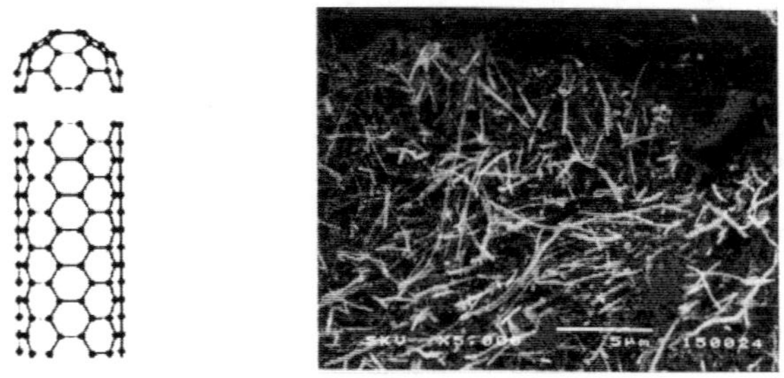

(a) CNT image. **(b)** SEM formed Al-CNT composite.

Figure 1. Schematic (a) CNT image and (b) SEM formed Al-CNT composite used this study.

This study was developed to determine (1) the high thermal conductivity of the Al-CNT material by adding the CF material and (2) the high radiation performance of the AC servomotor using the stator made by the Al-CNT composite.
The contents of this study are as follows:

- Evaluation of the thermal conductivity, mechanical strength, and thermal expansion,
- Thermal simulation of the AC servomotor using the Al-CNT stator, and
- Radiation evaluation of the AC servomotor using the Al-CNT composite to be applied in stator.

2. Fabrication and basic characteristic of the Al-CNT composite

2.1. Fabrication of the Al-CNT composite

The CNT made by Showa Denko was VGCF created using the CVD method, which at the same time created the carbonization and fiber formations, and the structure was SWCNT of approximately 200 nm in diameter and 200 μm in length. Figure 1(a) and 1(b) show a schematic CNT image and a SEM formed Al-CNT composite, respectively. The aspect ratio was about 10^4, and the outward appearance was like a black cotton. It was found that there was no affinity between melted Al and CNT. The formation process included CF lattice sheet to obtain affinity characteristics. Additionally, it was a necessity to have an anisotropic thermal conductivity of the Al-CNT composite to obtain the high performance of thermal radiation, as CNT has high anisotropic thermal conductivity of about 1000 W/(mK) in length. A fabrication process was performed by melting a mixture of granular Al of 200 μm and CNT at pressed atmosphere at same time. A fabricating condition and a formed Al-CNT composite are shown in Table 1 and Figure 2, respectively.

2.2. Thermal and mechanical characteristics of the CNT composite

The fabricated CNT rod composites were measured and evaluated for thermal conductivity in radius r and lengthwise z directions and mechanical breaking stress. As listed in Table 2, the sum total of Al and CNT was set to about 27 wt%. The evaluating instruments are as follows: thermal conductivity using
LFA457 (Netzsch) of sample size of about 10 mm in diameter and 3 mm in thickness, breaking stress using the Instron-type testing machine (Shimadzu Corporation), and coefficient of thermal expansion using Advance Riko, Inc. The test pieces were 5 mm in diameter and 20 mm in length.
A conventional Al using die casting for the motor stator has 92 W/(mK) thermal conductivity and 280 MPa breaking strength. However, the characteristic from the fabricated CNT composite was more than the conventional values. As shown in Table 2, the test piece of No. 2 indicated an excellent value of

169 W/(mK) in the radial direction and 173 W/(mK) in the lengthwise direction. Additionally, it was found that breaking strength increased in correspondence with increasing CF component. There is an attention point that the maximal thermal conductivity obtained 23 wt% CNT and other points have values less than it.

Table 1. Fabrication conditions of the Al-CNT composite.

Material	Al (ADC12)	Less than ϕ 0.2 (mm)
	CNT (Showa Denko)	200 nm diameter
		200 μm length
	CF (TORAY)	
Formed condition	Melting temperature	680 °C
	Formed pressure	9.8×10^5 N

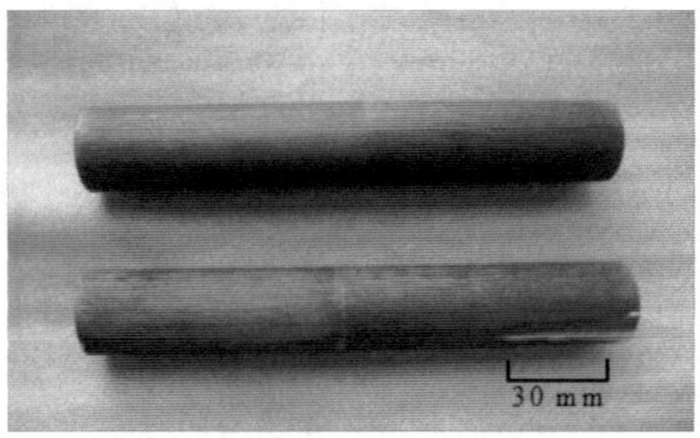

Figure 2. Formed conditions of the Al-CNT composite.

Table 2. Breaking stress and thermal conductivity with varied total CF and CNT composite.

No.	Component (wt%)	Breaking stress (MPa)	Thermal conductivity (W/(mK))
1	CF4-CNT23-Al	r direct. 450	r [a] direct. 124
		z 497	z [b] 125
2	CF14-CNT7-Al	r direct. 476	r direct. 169
		z 223	z 173
3	CF24-CNT3-Al	r direct. 871	r direct. 140
		z 714	z 153

[a] r is radious direction.
[b] z is lengthwise direction.

3. Thermal simulation of the motor

3.1. Analytical model and boundary condition

The thermal simulation of the AC servomotor with a square flange size of 100 mm used ANSYS thermal software (FEM) [6,7]. The analytical machine conditions and thermal conductivity of parts are shown in Table 3. An analysis model scale is selected in one-four size in the direction of circumference, and parts of the heating source are set to winding coils in a stator. The boundary condition assumed that the generated heat in coils radiated to the air through the stator and both blankets (front and back). The element of FEM used was a three-dimensional tetrahedron with 10 nodes.

The result of thermal conductivity under the condition of a half-symmetric structure about the motor that used Al (ADC12) frame is shown in Figure 3. It was set that a quantity of heat from the coils was 85 W/m^3, the external temperature of stator flame was 15 °C, and the thermal conductivity of the frame was 92 W/(mK). As a result of the steady-state analysis, the temperature of coils and surfaces of the stator flame was 122 °C and 60 °C, respectively. Therefore, it was considered that the coil temperature increased to 107 °C at the rate operating condition of motor with current of 5.7 A on AC 200 V and the temperature of the stator increased to 45 °C.

Table 3. Thermal analysis condition of the FEM machine and thermal conductivity of parts.

Item	Condition / Value
Simulating machine	Intel Celeron CPU, 600 (MHz)
Time	8 (min)
Solver	ICCG method, $\delta A/A$<1.0e $^{-7}$
Number of node	122,073
Number of element (Tetrahedron with 10 nodes)	85,197
Parts (Material)	**Thermal conductivity (W/(mK))**
Rotor of motor (Fe)	83.6
Gap, Space (Air)	0.0241
Front/Rear blanket (Al)	92.0
Stator frame (Al)	92.0
(Al-CNT)	36.0 - 232
Coil (Cu)	403
Molding of stator (Epoxy)	0.95

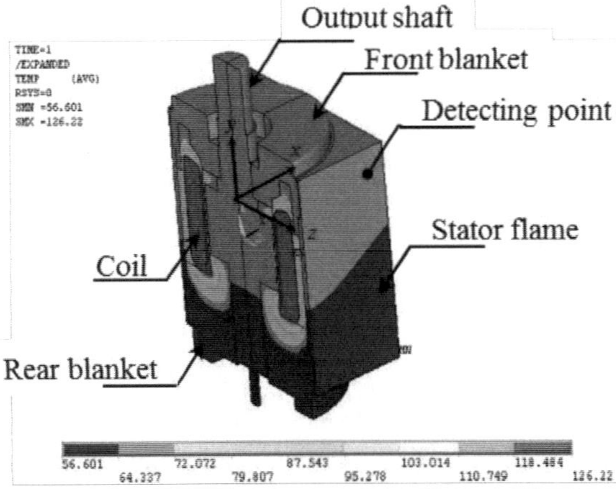

Figure 3. Temperature contour of the AC servomotor with Al frame.

3.2. Simulation of thermal conductivity on the motor that used CNT composite

The conventional thermal conductivity using Al stator in the motor frame was 92 W/(mK). However, in this study, the thermal conductivity consisting of Al-CNT composite was 171 W/(mK). Therefore, the thermal conductivity simulation in the motor that applied the CNT composite was tested. Figure 4 shows the simulation results of temperature dependence on the thermal conductivity of stator and coil using the CNT composite. It was shown that the stator temperature increased but the coil temperature decreased with increasing estimated thermal conductivity of stator. This is the reason that coil temperature decreased as thermal conductivity increased in the stator. A caution feature result is that temperature increase appears in the stator when thermal radiation reaches saturation. It was found that the temperature increase of stator is possible to decrease using that of the CNT composite. In other words, more than the rate current value. According to the FEM simulation result, the coil temperature is 114°C, which is less than 13 °C of the conventional Al stator using Al-CNT 14 wt% composite with thermal conductivity of 171 W/(mK).

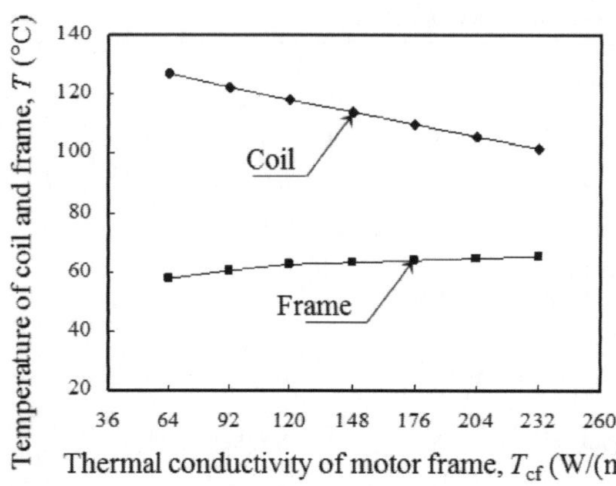

Figure 4. Temperature dependence on thermal conductivity of the frame using the FEM.

4. Fabrication and evaluation of the CNT motor

4.1. Basic characteristic and fabrication of the CNT motor

Two kinds of motors made using Al and CNT composite were evaluated to confirm the effect of thermal conductivity of stators. The common specifications of the two kinds of motors are shown in Table 4. Both motors have a square flange size of 100 mm, length of 120 mm, and weight of about 4.5 kg. As shown in Figure 3, the temperature of the motor was measured at point D_p using C-A thermocouple (B13-440). An AC servomotor made by the CNT composite stator is shown in Figure 5. As the surface of CNT composite stator had rough stripe patterns, the external surface obtained smooth surface using black color coating. Then, the measurement equipment of motor characteristics used was an optical encoder with rotary speed and acceleration at the opposite position of output power axis, and the motor controller used was a servomotor unit of TA8110 made by Tamagawa Seiki Co.

The dependence torque on rotary speed of both motors obtained linear characteristics between rotary speed of 0 to 4,510 rpm and torque of 10.2 Nm to 0, respectively. The dependence torque on the current of both motors obtained torque constants of 0.59 Nm/A under the condition of rate current of 5.6 A and torque of 3.3 Nm. The motor fabrication using the CNT composite led to the preparation of the radiation test.

Table 4. Specifications of the servomotor with Al and Al-CNT frames.

Item	Value
Output power, P_o	1.0 (kW)
Rated torque, T_r	3.3 (Nm)
Peak torque, T_p	9.9 (Nm)
Rated current, I_r	5.7 (A)
Rated speed, S_r	3,000 (rpm)
Max. speed, S_m	4,500 (rpm)

IOP Conf. Series: Materials Science and Engineering **311** (2018) 012003 doi:10.1088/1757-899X/311/1/012003

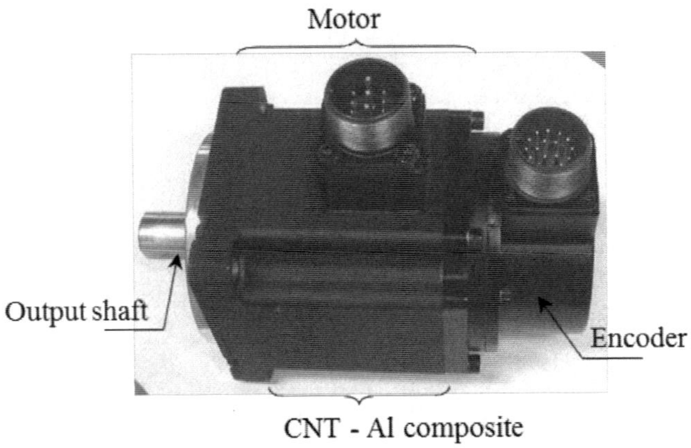

Figure 5. AC servomotor using the stator frame with the Al-CNT composite.

4.2. Radiation characteristic of the CNT composite motor

The temperature of the external atmosphere and the stator was measured at the same operating conditions for both motors. A change in temperature was observed until steady state at measurement points. An end time was decided by making sure it is less than 1.4 °C per 30 min of temperature decrease. Because direct temperature measurement of the coil would not be executed as coil was set at inner stator, the temperature was estimated by varied coil resistivity (i.e., the rise-of-resistance method) [8].

The increasing temperature dependence on the two kinds of motors at three points under the condition of rotational speed with 3,000 rpm and load torque with 3.3 Nm is shown in Table 5. The coil temperature was decreased to 19 °C at the same time as the stator frame temperature was increased to 5 °C in the case when the stator used the CNT composite. The result shows that the input current of the CNT motor can increase until it is equivalent to a value with coil temperature decrease. It fairly well confirmed the radiation effect of the applied CNT composite. The design and measurement values of the motor with CNT stator are shown in Table 6. The consisted material of CNT composite motor was the same as the conventional one without stator material. As a result, it was obtained that the CNT motor improved the decrease of coil temperature and radiation characteristics.

According to the obtained temperature characteristics, the composite motor using CNT decreased consumption energy by 16 % with respect to the conventional one. This fact means that it is possible to increase the rate current to 6.8 %. In other words, the rate current can be from 6.25 to 5.6 A and 3.7 to 3.5 Nm in the case of rate torque.

Table 5. Increasing temperature dependence of two motors.

Frame material	Measurement position	Detected temperature T_d (°C)	Atmosphere temperature T_a (°C)	Increasing temperature T_i(°C)
Al	Frame surface	57.0	15.6	41.4
	Coil	122.8	15.6	107.2
Al-CNT	Frame surface	62.0	18.0	44.0
	Coil	103.5	18.0	85.5

IOP Conf. Series: Materials Science and Engineering **311** (2018) 012003 doi:10.1088/1757-899X/311/1/012003

Table 6. Design and measurement values of the motor Al-CNT stator.

Output power, P (W)	1.0 k	1.0 k
Rated torque, T_r (Nm)	3.5	3.7
Peak torque, T_m (Nm)	10.1	10.3
Rated current, I_a (A)	5.9	6.3
Rated speed, S_r (rpm)	3.0 k	3.0 k
Max. speed, S_m (rpm)	4.5 k	4.5 k

4.3. Studies in the future

The development of the CNT composite and its application in motor were performed at the same time using thermal FEM. As a result, the highly efficient motor has improved the radiation characteristic using the Al-CNT composite stator. However, the future problems are as follows: (1) the cost of the CNT material, (2) confirmation of the influence on the human body, and (3) no good affinity of CNT for Al.

5. Conclusions

The nanotechnology applied in motors expects a bigger revolution in the future, which is why CNT using high thermal conductivity is particularly important. In this study, the making of the CNT composite and high thermal radiation motor was studied.

The obtained results are as follows;

- Successful fabrication of Al 25 wt%-CNT composite; the CNT composite had twice thermal conductivity compared to Al (ADC12);
- A novel method of increasing the affinity between Al and CNT using CF; and
- Successful thermal radiation improvement of the motor stator, which then could increase to 6.8 % motor current at the same size.

6. References

[1] Yhachi S and Shunji B Basic of carbon nano-tube 2001 Japan *Corona Publish Co.* 58.

[2] Kunio N Production technology and applied development of carbon nano-tubes 2004 *Nanotech college seminar text.*

[3] Takashi I and Yutaka I and shin I and shunji S Composite of distributed fiber with carbon nano-fiber *Japan Patent 2004-131538.*

[4] Showa Denko Data *sheet of product safety* 2001.

[5] Riichiro Saito Physical properties of carbon nano-tubes 2001 *Imperial College Press* UK 35.

[6] ANSYS INC. Introduction of FEM 2000 *DN-S231J* 17.

[7] ANSYS INC. Introduction of technology analysis 2001 *Rikougakusha Publishing Co Ltd.* Pp.46-52.

[8] Richard E and William M Handbook of electric motors 1995 *Marcel Dekker Inc.* pp.416-417.

Acknowledgments

The authors would like to acknowledge the advice and assistance of Prof. Endo and Coordinator Morimoto of Shinshu University, respectively.

IOP Conf. Series: Materials Science and Engineering **311** (2018) 012004 doi:10.1088/1757-899X/311/1/012004

Smart reconfigurable parabolic space antenna for variable electromagnetic patterns

Sahil Kalra[1], Rituparna Datta[2], B.S.Munjal[3], and Bishakh Bhattacharya[4]

[1,4]Department of Mechanical Engineering, Indian Institute of Technology Kanpur, India
[2]Graduate School of knowledge Service Engineering, KAIST, South Korea
[3]Space Applications Centre, Indian space research organization, Ahmedabad, India
bishakh@iitk.ac.in

Abstract. An application of reconfigurable parabolic space antenna for satellite is discussed in this paper. The present study focuses on shape morphing of flexible parabolic antenna actuated with Shape Memory Alloy (SMA) wires. The antenna is able to transmit the signals to the desired footprint on earth with a desired gain value. SMA wire based actuation with a locking device is developed for a precise control of Antenna shape. The locking device is efficient to hold the structure in deformed configuration during power cutoff from the system. The maximum controllable deflection at any point using such actuation system is about 25mm with a precision of $\pm 100\mu$m. In order to control the shape of the antenna in a closed feedback loop, a Proportional, Integral and Derivative (PID) based controller is developed using LabVIEW (NI) and experiments are performed. Numerical modeling and analysis of the structure is carried out using finite element software ABAQUS. For data reduction and fast computation, stiffness matrix generated by ABAQUS is condensed by Guyan Reduction technique and shape optimization is performed using Non-dominated Sorting Genetic Algorithm (NSGA-II). The matching in comparative study between numerical and experimental set-up shows efficacy of our method. Thereafter, Electro-Magnetic (EM) simulations of the deformed shape is carried out using electromagnetic field simulation, High Frequency Structure Simulator (HFSS). The proposed design is envisaged to be very effective for multipurpose application of satellite system in the future missions of Indian Space Research Organization (ISRO).

1. Introduction

Antennas are the basic constituents of a satellite communication system. It acts as a transferring medium of radio waves between two ground stations commonly known as uplink and downlink. Depending upon frequency band, polarization, field of application and mode of operation, antennas can be categorized into various groups and sub groups as shown in Fig. 1. Each sub group in Fig. 1 has its own type depending upon the characteristics of service and functionality. Among this group, parabolic reflectors are most common for space applications.

Using the concepts of Geometric Optics (GO) it can easily be proved that for a perfect parabolic surface, all rays covers equal distance from focus to the aperture plane. Hence, far field radiations pattern of such reflector produces a similar pencil beam along focus of the parabola. However, the bandwidth (HPBW) at variable frequencies / wavelength shows the difference. The band width is inversely proportional to the operational frequency. For example, increasing the frequency by 10% reduces the HPBW by the same amount as shown in Fig.2. One of the categories of parabolic reflectors are shaped reflectors which have relatively better HPBW. However, their shape is unique corresponding to the geographical area to scan and the gain in given direction is also fixed. Hence, for

space applications there is high demand of a reflector whose gain and footprints can actively be steered and shaped.
In the previous studies reconfigurable parabolic antennas actuated using Piezoelectric actuators were studied by researchers [3, 4]. Most of the past studies were performed by considering assumption which cannot be verified experimentally [1-2] especially in the space environment. However, the space environment conditions require non-degradable actuators with low power consumption.
In order to overcome these challenges, we have demonstrated a Shape Memory Alloy (SMA) actuated single feed antenna system which can change the gain of radiation pattern and actively change the footprints. A C-band parabolic reflector made of flexible polycarbonate Lexan material (E = 1.5GPa) (shown in Fig.12) is presented for the analysis. An attempt has been made to present a unique concept for the first time as a dedicated locking device which is based upon the requirement to lock the shape of antenna even though power supply is cut off from the system [5]. As continuous power supply from spacecraft may destabilize the satellite in Roll, Pitch & Yaw axes, hence the need has been felt for such innovation. The effect of shape change of antenna on EM pattern is also studied in this paper. Thereafter, we have presented a multi-objective Genetic Algorithm (GA) based approach to minimize the voltage requirements and minimizing the error between the desired and achieved shape.

Frequency Basis: 1. VLF (3-30KHz) 2. LF (30-300KHz) 3. HF (3-30MHz) 4. VHF (30-300MHz) 5. UHF (300-3000 MHz) 6. SHF (3-30GHz) 7. EHF (30-300GHz)	**Aperture Antennas:** 1. Wire Antennas 2. Horn Antennas 3. Parabolic reflective Antenna 4. Cassegrain Antenna
Polarization Basis: 1. Linearly Polarized Antenna 2. Circularly Polarized Antenna	**Radiation Pattern Basis:** 1. Isotropic Antenna 2. Omnidirectional Antenna 3. Directional Antenna 4. Hemispherical Antenna

VLF: Very Low Frequency; LF: Low Frequency; HF: High Frequency; VHF: Very High Frequency: UHF: Ultra High Frequency; SHF: Super High Frequency; EHF: Extremely High Frequency

Fig. 1. Various Categories of Antenna

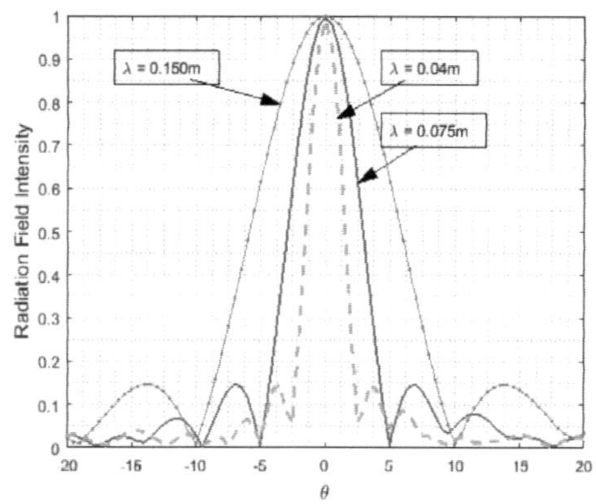

Fig. 2. Variation of bandwidth at different frequencies

2. Structural shape analysis of paraboloid shell

Finite Element (FE) analysis for the structure is performed using ABAQUS software. The parabolic geometry is meshed with 4-noded shell elements having six Degrees of Freedom (DoFs) at each node. Total number of 402 nodes and 373 elements are considered to perfectly model the deformation of the system and stress profile. It is analyzed that with a mesh of 402 elements, the degree of stress discontinuity between the adjacent nodes decreases significantly. These nodes are subdivided into 3 major sets namely (i). Actuated node set (ii) Fixed node set and (iii) Internal node set. Vertical forces of different magnitude are applied on various actuated nodes and deformed shape of antenna is analyzed. With the application of force at a Control Pont (CP) (see Fig.3), the inflation and deflation on antenna surface is shown in Fig.4. This shows the coupled behavior of a nodal position with neighboring nodes. Hence, in order to perform shape control the following procedure is adopted.
The undeformed nodal coordinates, stiffness matrix (K) containing material information of each node are exported from the FE software. After applying the fixed boundary conditions, the stiffness matrix is statically condensed using the Guyan reduction technique. For DoFs other than y-direction, all internal nodes and fixed nodes are eliminated. Subsequently, a direct mathematical relation between applied force (F), initial coordinates (x_{ini}) and deformed coordinates (x_{def}) is established as shown in equation (1).

$$F = K(x_{ini} - x_{def}) \quad (1)$$

2.1. Selection of SMA for shape control of antenna structure

In order to select an appropriate SMA actuator for the reflector, the stiffness contribution at individual CP of structure is measured by loading the CP using dead weights and measuring the deflection by the single point laser sensor. The force to deflection curve is shown in Fig.5, the ratio K (400N/m) is linear upto the deflection range of approximately 25mm. No major difference in the stiffness value at 8 CPs was recorded, as the structure was symmetric. This range of deflection is considered as elastic range. Based upon this stiffness value, heating and cooling force requirement, maximum displacement requirement on a CP, SMA wire of diameter 375μm was chosen.

2.2 Analysis of SMA actuator with constraint structure

For this application SMA wire is heated using Joule effect. Hence, a simple mathematical model is developed in which SMA is considered as a lumped body [6]. When electric current flows through it, it performs the mechanical work in addition to heat losses. As current flow stops it starts heat rejection. As the total energy of the system has to be conserved so one can equate the total electrical energy consumed by the actuator to work performed with some heat losses as shown in equation (2).

Since, the magnitude of heat losses due to radiation and conduction, latent heat and mechanical work is small (20%) in comparison to the other factors, hence equation (2) is simplified to equation (3). For multiple voltage inputs the analytical solution of (3) is given in equation (4). Where, T_a is the ambient temperature, t_F is the end time, t_I is the initial time period, A_i is the voltage amplitudes, t_i^e is the end time of voltage A_i, t_i^s is the start time of voltage A_i.

$$\underbrace{\frac{V^2(t)}{R}}_{\substack{Power\\supplied}} = \underbrace{hA(T(t)-T_a)}_{\substack{Convective\ heat\\loss}} + \underbrace{\sigma_{rad}A(T^4(t)-T_a^4)}_{Radiative\ heat\ loss} + \underbrace{\rho V c_p \frac{dT}{dt}}_{\substack{Specific\\heat}} + \underbrace{kA\frac{(T(t)-T_a)}{l}}_{\substack{conductive\ heat\\loss}} + \underbrace{\rho V \delta H \frac{d\xi}{dT}}_{\substack{Latent\\heat}} + \underbrace{\frac{dW_{mech}}{dt}}_{\substack{Mechanical\\work}} \quad (2)$$

$$\frac{V^2(t)}{R} = hA(T(t)-T_a) + \rho V c_p \frac{dT}{dt} \quad (3)$$

$$T = T_a + \frac{1}{e^{t_1 t_F} - e^{t_1 t_I}} \sum_i^n \frac{A_i^2}{\tau_1 \tau_2}\left(e^{\tau_1 t_i^e} - e^{\tau_1 t_i^s}\right) \quad (4)$$

where, $\tau_1 = \frac{\rho V c_p}{h_c A}, \tau_2 = \rho V c_p R_w$

Using Brinson model [7], the displacement at a CP is analyzed by incorporating the stiffness (K) in equation (5). The strain profile of a CP on the constraint structure actuated using SMA wire (diameter = 375μm) by applying voltage profile is shown in Fig.6. A direct mathematical relationship between the displacement of CP and temperatures of SMA wire is developed using rational polynomial algorithm in MATLAB. The relationship for heating cycle is shown in equation (6). This relationship covers the temperature range from 30°C to 100°C. Cooling cycle of SMA is not considered due to the presence of locking device that is discussed in section (Experimental setup).

$$\sigma - \sigma_0 = D(\xi)\epsilon - D(\xi_0)\epsilon_0 + \Omega(\xi_0)\xi_{s_0} + \Theta(T - T_0) + \frac{kx}{A} + \frac{kx_0}{A} \quad (5)$$

$$x = \frac{p_1 T^4 + p_2 T^3 + p_3 T^2 + p_4 T + p_5}{T^2 + q_1 T + q_2} \quad (6)$$

$p_1 = 5.824e^{-7}; p_2 = -0.0003374; p_3 = 0.04289; p_5 = 26.84; q_1 = -154.6;$ $q_2 = 6117$

3. Shape optimization

The shape optimization of the antenna is formulated as a multi-objective optimization problem where, the optimization is performed by considering the eight CPs on the structure. The following objective functions are considered:

- Minimization of mean square error between the desired and current shape

IOP Conf. Series: Materials Science and Engineering **311** (2018) 012004 doi:10.1088/1757-899X/311/1/012004

- Minimum total voltage requirement to actuate the actuators

With a constraint that stress and strain in wire should not increase the safe stress of 180 MPa and 0.05% respectively. At the safe stress level effect of functional fatique can be neglected.

$SE = \frac{\epsilon^2 E}{2}$, strain ($\epsilon$) at temperature (T) where, $E = E_A + \xi(E_M - E_A)$

Displacement error = $x_{desired} - x$,

Mean square error, (MSE)= $\sqrt{x_{d_i}^2 - x_i^2}$

Power, $\frac{V^2}{R}$; $R = R_A + \xi(R_A - R_M)$

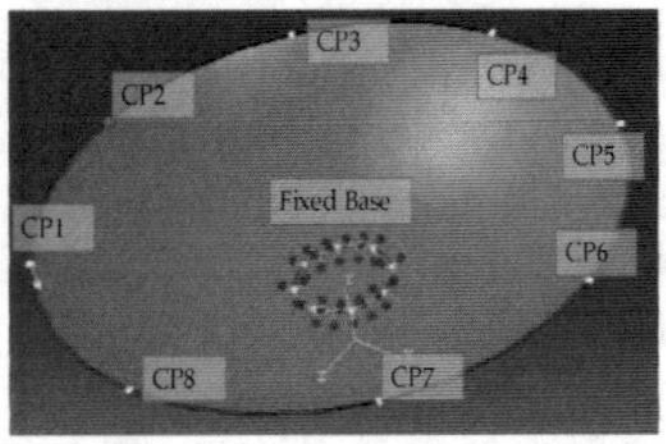

Fig.3. Eight Control Points on the structure

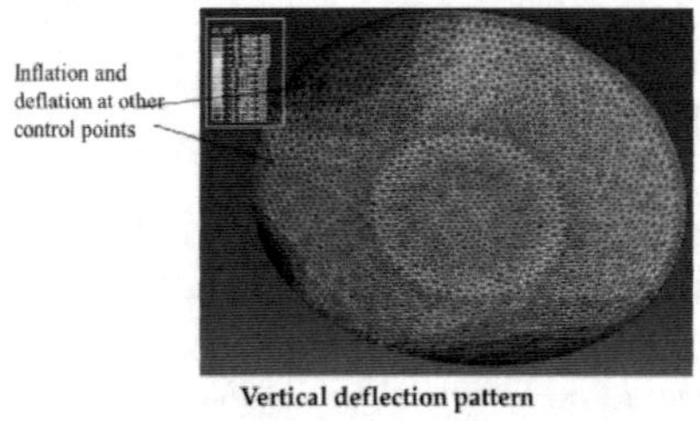

Fig.4. Inflation and deflation on structure while deforming a CP

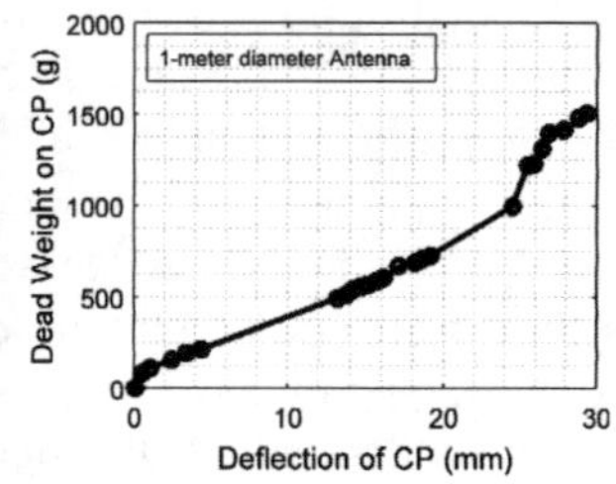

Fig.5. Experimental Result to calculate the stiffness at a CP

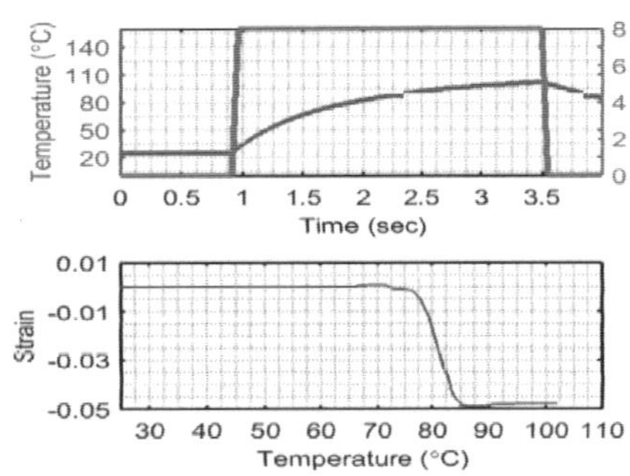

Fig.6. Displacement of a CP by heating SMA

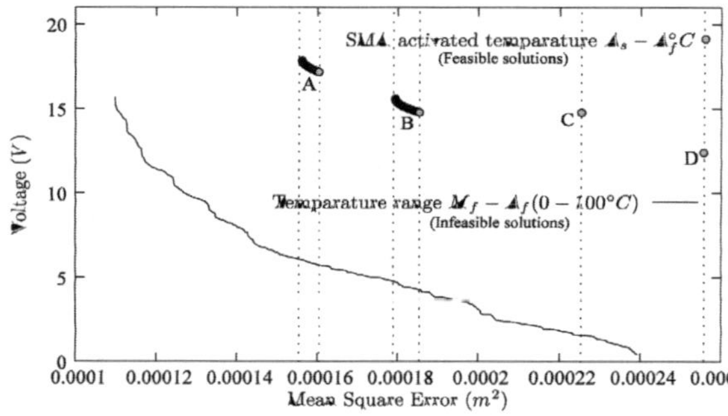

Fig.7. Non-dominated solutions between Mean Square Error versus Voltage for Case-I

The multi-objective optimization problem is solved using evolutionary multi-objective optimization. Constrained Non-dominated Sorting Genetic Algorithm-II (NSGA-II) [8] is used to solve the multi-objective optimization problem. The following parameters for are considered:

Population size = 40n, (n=Number of variables)
Number of Generations = 500,
SBX probability= 0.9,
SBX index = 10,
Polynomial mutation probability = 1/n, and
Mutation index = 100

Two different cases of the parabolic space antenna are considered commonly known as steering and shaping. For Case-I, a desired displacement array of (250, -120, -100, 250, -100, 200, -60, 140)µm and for case-II desired array displacement of (250, -50, -50, 0, 200, 0, -50, -80)µm at CP1-8 is considered. The non-dominated solutions between mean square error versus voltage is shown in Fig.7. The shape of the Pareto-front is non-convex. Due to the presence of non-linearity classical weighted sum approach is unable to generate non-dominated solutions [9]. The figure clearly shows that a better set of non-dominated solutions (both in terms of mean square and voltage) can be obtained for the temperature range between M_s to A_f i.e between (52−100°C). However, the actuating temperature of

SMA is between A_s to A_f (70−100°C) including various constraint recovery stresses. This is the reason for the infeasibility of all those solutions ranging between M_f to A_f . All those solutions having temperature below 70°C will not be actuated. As a result, the feasible range for all temperatures must lie between 70−100°C. The figure also shows that the range of mean square error is between 156 µm to 256µm and the voltage ranges from 12.39 to 17.86V. For minimum mean square solution temperature at first actuator should be 92.74°C, temperature at the fourth actuator should be 70°C and all other temperature must be zero. On the other hand, for minimum voltage solution all the temperature must be zero. The meaning is that there will not be any movement of the antenna.

The change in temperature range will change the non-dominated solutions. Multi-objective optimization method works with two different spaces i.e. variables space and objective space. The temperature range between (A_s to A_f) has reduced the feasible region in the variable space which has been restricted the objective space. The feasible solutions has four different zones shown with A, B, C and D. All solutions in these zones have equal importance.

The range of mean square error is ranging from 37µm to 145µm whereas minimum voltage is 12.39V and maximum voltage is 18V. Temperature at first actuator should be 97°C, sixth actuator temperature should be 70°C and all other actuator temperature should be zero to obtain minimum mean squared error solution. However, all the temperature must be zero to achieve minimum voltage solution. If a user wants to use minimum value of mean square error, the left most solution in zone A must be chosen. On the other hand for lowest voltage solution D will be choice. We will analyse the effect of temperature in each objective function in next subsection.

The Pareto-optimal solutions for case-II can be seen in Fig.8. A better mean squared error solutions has been obtained in Case-II as compared to Case-I. Case-II also contains four zones like Case-I. The comparison between case-I and Case-II has been shown in Fig.9. The figure clearly shows that the obtained solutions are better in both objective functions in Case-II than Case-I.

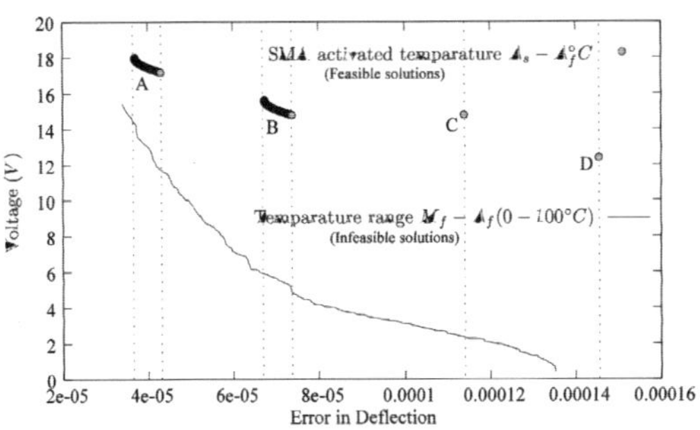

Fig.8. Non-dominated solutions between Mean Square Error versus Voltage for Case-II

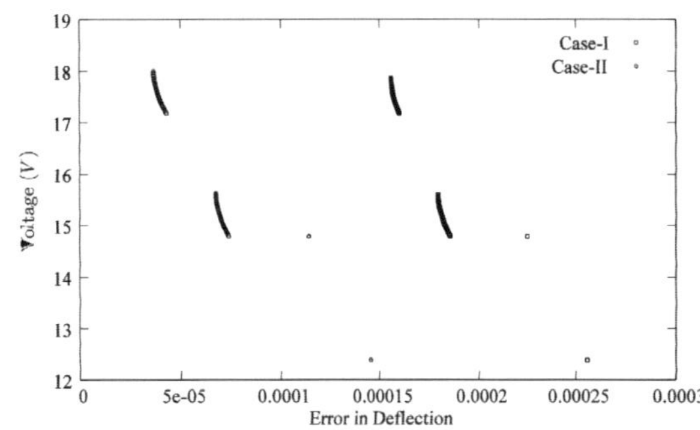

Fig.9. Comparison of non-dominated solutions between Mean Square Error versus Voltage for both cases

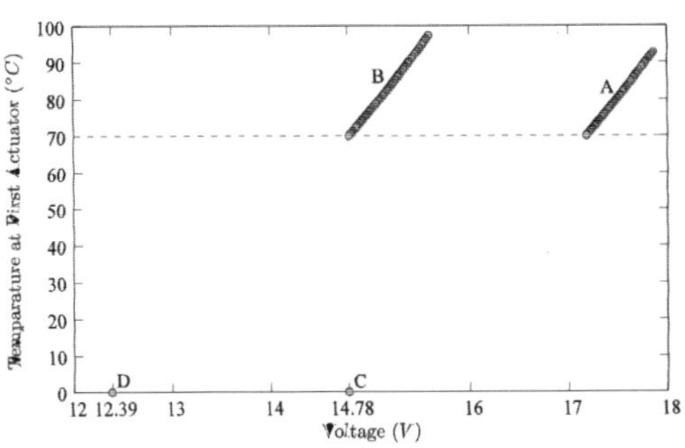

Fig.10. Relationship between Voltage with temperature of First Actuator for Case-I

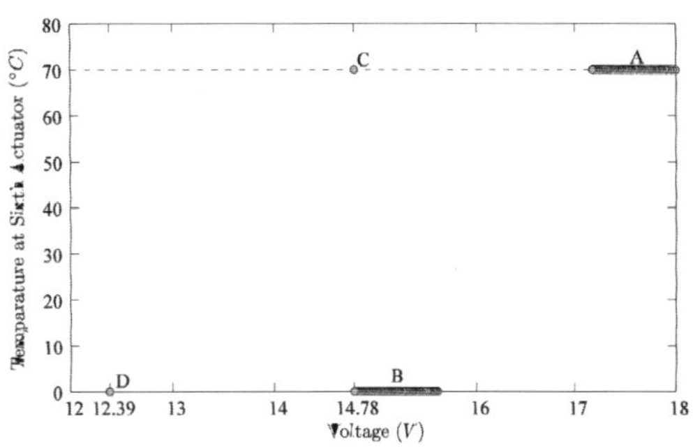

Fig.11. Relationship between Voltage with temperature of Sixth Actuator for Case-II

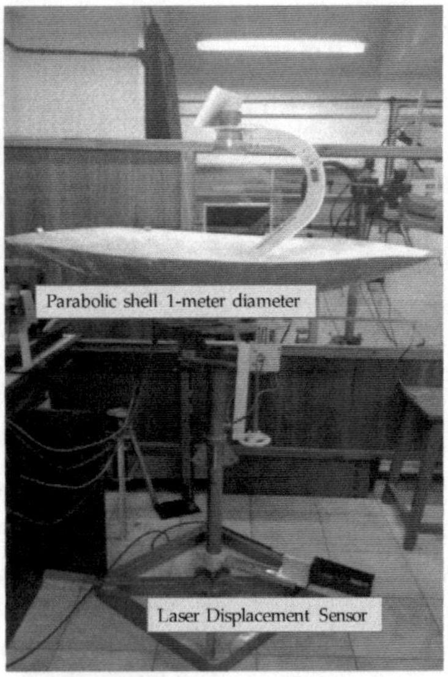

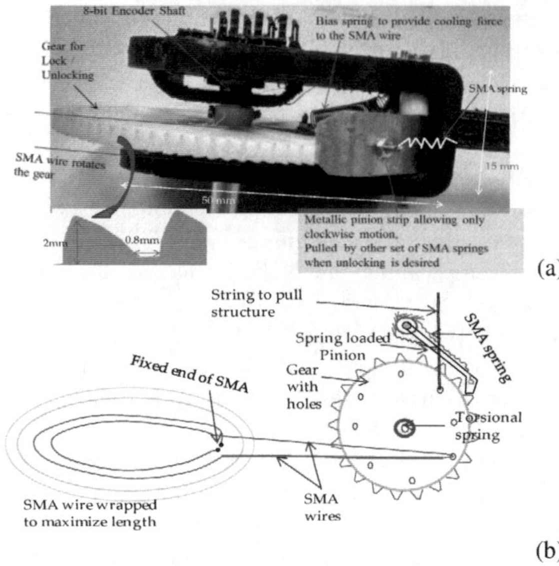

Fig.12. Experimental setup of Reconfigurable parabolic antenna

Fig.13 (a). Locking device (b) A schematic of SMA wire wrap around antenna with locking device working demonstration

3.1 Temperature effect in objective functions

Here, we analysed the effect of temperature in both the objective functions i.e. mean square error and voltage. As mean square error and voltage are conflicting to each other, the relationship with any one of the objective functions with temperature is sufficient. The other will follow inverse relationship. Here, we use voltage to establish a relationship between temperatures of different actuators.

Fig.10 shows the relationship between temperatures of the first actuator versus voltage. Here we can see four different relationships exist. These four zones have been marked with A, B, C and D. These four zones are nothing but four zones of non-dominated solution shown in Fig.7. In A and B zone voltage is direct proportional to temperature which means that if we want an increase in voltage temperature must be increased and vice versa. For zones C and D temperature should be kept at ambient temperature (T_a).

Similar to the first actuator, the relationships of other actuators is developed. A user has to keep the temperature at 70°C to achieve a solution in A and C zones. On the other hand, to achieve voltage at zones B and D, the temperature must be kept at T_a. The relationships between voltage and actuator temperature for Case-II is shown is Fig.11. User can vary the temperature to obtain voltage accordingly. This relationship will help the decision maker to take decision by varying the temperature based on desired voltage.

4. Experimental setup

The parabolic reflector setup with single feed mount is shown in Fig.12. The CPs are actuated using auto locking device as shown in Fig.13(a). One end of the SMA wire is fixed with the base of the reflector and the other end rotates the geared pulley mounted at the fixed platform as shown in Fig. 13(b). The working principle of the locking mechanism developed in this work is similar to the ratchet mechanism in which a metallic strip (pinion) being pulled by bias spring that allows only anti-clockwise rotation of the pulley. When unlocking is desired, the pinion can be pulled using SMA spring. In order to keep the shaft rotation frictionless, roller bearings are used. For sensing the shaft rotation, an 8 bit absolute encoder with 128 pulse per revolution is mounted with the shaft of pulley. The digital data are fed into the controller, which using a PID control algorithm decides the power supply required to the SMA actuators. The block diagram of the closed loop controller is shown in Fig.14.

4.1 EM Simulation results

The deformed mesh from ABAQUS is converted to surface thereafter EM software package High Frequency Structure Simulator (HFSS) and MATLAB are exclusively used to generate the EM patterns. The results of radiation pattern for deformed and un-deformed Antenna geometry is shown in Fig.15, 16. The results clearly show the shifting of major lobe of EM pattern. This shifting can be very useful for the EM analysts. The following advantages can be undertaken using the shift:

- Two beam modes, steering and shaping can be predominately achieved using this system. Hence, in space by using reconfigurable antennas system bigger landmasses, different landmasses can be covered.
- Recovery of manufacturing errors, in orbit thermal distortion, loosed structural joints, material degradation, creep can be compensated with this technology.
- The precise antenna surface control for beam broadening applications, can be easily achievable using reconfigurable antenna. The major advantage of such systems is in the field of radio astronomy.
- With similar system researchers at Ohio state University [3] successfully demonstrated beam steer from the North America to South America.

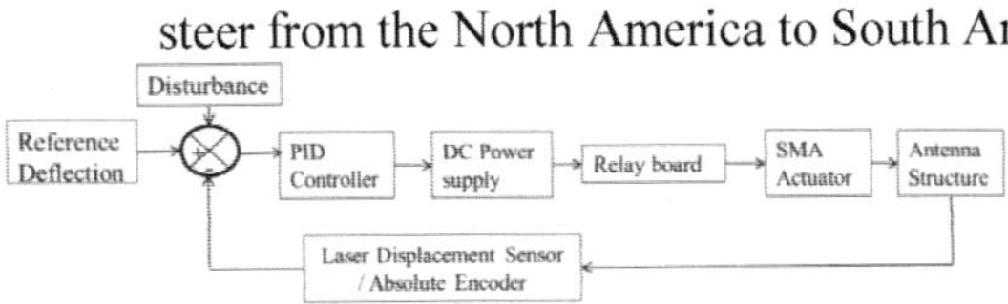

Fig.14. Closed Feedback control loop for Antenna

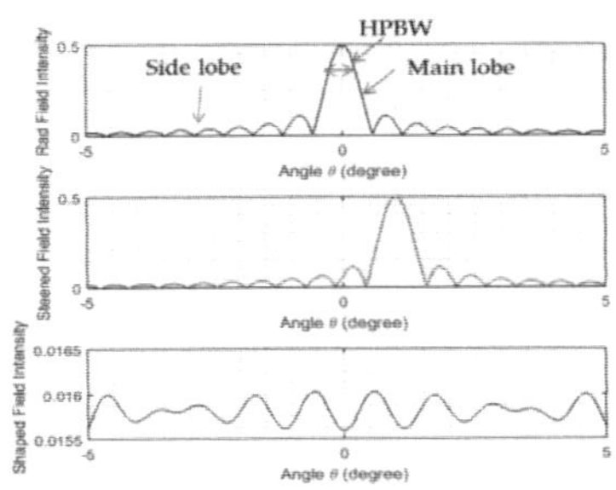

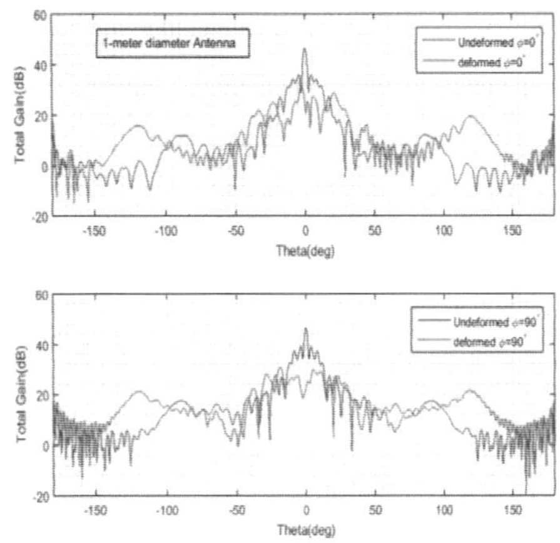

Fig.16 Far field intensity for the analysis of deformed Antenna using MATLAB

Fig.15 EM pattern for un-deformed and deformed geometry at $\phi = 0{,}90^o$ using HFSS

5. Conclusion and future scope

The feasibility of shape control of RF reflector using SMA based actuation system is demonstrated in this paper. The steering and shaping modes of EM pattern can be useful in space for coverage of larger or different landmasses at the same time. As a Proof-of-Concept model, the present antenna structure has to be verified under Compact Antenna Test Facility (CATF) of ISRO, so that the Design Validation can be established for space worthy hardware in near future. This approach has potential applications in ground segment and Air Borne systems as well, which are being explored by the authors.

6. References

[1] Landeros S., Neri R., Samano R., *A Tutorial on the Synthesis of Single Shaped Reflectors in C, KU, and KA Bands* Taylor and Francis 26(2), p. 131–154, (2006)

[2] Cherrette, A. R., Lee S W., Acosta R J., *A Method for Producing a Shaped Contour Radiation Pattern Using a Single Shaped Reflector and a Single Feed*, IEEE Transactions on Antennas and Propagation 37(6), p. 698-706, (1989)

[3] H. Yoon, Design, Modeling, and Optimization of a Mechanically Reconfigurable Smart Reflector Antenna System, PhD dissertation, The Ohio State University, (2002)

[4] Chawla, A., Bhattacharya, B., and Munjal, B. S., *Optimal Actuation of SMA-Wire Network for*

Adaptive Shape Control of a Space Antenna System, Proceedings in IEEE, ICARCV, Singapore, (2014)

[5] Kalra, S., Bhattacharya, B., and Munjal, B. S., *Design of shape memory alloy actuated intelligent parabolic antenna for space applications*, Smart Material and Structures, IOP Science, vol.26(9) (2017)

[6] Meier, H. and Oelschlaeger, L., *Numerical thermomechanical modelling of shape memory alloy wires*, Materials Science and Engineering A, Elsevier vol. 378, p. 484–489, (2004)

[7] Brinson, L. C., One Dimensional Constitutive Behavior of Shape Memory Alloys: Thermomechanical Derivation With Non-Constant Material Functions and Redefined Martensite Internal Variable, Journal of intelligent material systems and structures, vol. 4, no. 2, p. 229–242, (1993).

[8] Deb, K., Pratap, A., Agarwal, S. and Meyarivan, T., *A fast and elitist multiobjective genetic algorithm: NSGA-II*, IEEE Transactions on Evolutionary Computation, vol. 6, no. 2, p. 182–197, (2002)

[9] Deb, K., *Multi-Objective Optimization using Evolutionary Algorithms*, handbook, John Wiley & Sons, ISBN: 978-0-471-87339-6, (2001)

IOP Conf. Series: Materials Science and Engineering **311** (2018) 012005 doi:10.1088/1757-899X/311/1/012005

High-Performance AC Power Source by Applying Robust Stability Control Technology for Precision Material Machining

En-Chih Chang

Department of Electrical Engineering, I-Shou University, Taiwan, R.O.C.
E-mail: enchihchang@isu.edu.tw

Abstract. This paper presents a high-performance AC power source by applying robust stability control technology for precision material machining (PMM). The proposed technology associates the benefits of finite-time convergent sliding function (FTCSF) and firefly optimization algorithm (FOA). The FTCSF maintains the robustness of conventional sliding mode, and simultaneously speeds up the convergence speed of the system state. Unfortunately, when a highly nonlinear loading is applied, the chatter will occur. The chatter results in high total harmonic distortion (THD) output voltage of AC power source, and even deteriorates the stability of PMM. The FOA is therefore used to remove the chatter, and the FTCSF still preserves finite system-state convergence time. By combining FTCSF with FOA, the AC power source of PMM can yield good steady-state and transient performance. Experimental results are performed in support of the proposed technology.

1. Introduction

The AC power sources are broadly applied in precision material machining (PMM), such as high strength aluminum alloys, brass alloys, and stainless steel alloys [1]. The requirements of a high-performance AC power source must have low THD output-voltage, fast transient response and steady-state errors as small as possible in the face of different loading. To achieve such requirements, proportional–integral–derivative (PID) controller is usually used. However, when the system using PID controller under the case of a variable load rather than the nominal ones, cannot obtain fast and stable output voltage response [2]. Numerous control technologies are also reported in literature, such as wavelet transform technique, deadbeat control, repetitive control, and so on [3], [4], [5]. But, they are difficult to implement and have complex algorithms. Sliding mode control (SMC) with insensitivity to system uncertainties was first proposed in early 1950's. The SMC of AC power sources are also frequently used, but they adopt linear sliding function with slow convergence time [6]. In order to speed up the convergence speed, a finite-time convergent sliding function (FTCSF) is employed, and the system state can be driven to the origin within finite time [7], [8]. It is worth noting that the chatter will occur around the FTCSF once a severe nonlinear loading is applied. In the AC power source output, the high THD caused by the chatter easily yields, thus worsening the stability of the PMM. Firefly optimization algorithm (FOA) is a meta-heuristic technology inspired by the sparkling behavior of fireflies, and has been broadly used for solving optimal control problems in various engineering fields [9], [10]. The FOA is thus used to remove the chatter while the FTCSF is employed to speed up the convergence speed of the system state. By the combination of FTCSF and FOA, a closed-loop AC power source of PMM will reveal excellent performance, such as low total harmonic distortion of output voltage, fast transient response, and the reduction of the chatter.

Experimental results are given to prove the effectiveness of the proposed robust stability control technology. Because the proposed control technology is easier to realize, faster to converge, more effective to remove the chatter than prior methods, this paper will be a useful reference to correlative material researchers.

2. Proposed robust stability control technology for AC power source of PMM

Figure 1 displays the circuitry of AC power source converter of PMM, and the PWM inverter is a key unit in the controlled system. In case the output voltage v_o follows a sinusoidal reference voltage v_d exactly, the error between v_o and v_d can be converged to the origin. By defining $x_{e1} = v_o - v_d$ and $x_{e2} = \dot{x}_{e1}$, the error state-space matrix is represented as

$$\begin{bmatrix} \dot{x}_{e1} \\ \dot{x}_{e2} \end{bmatrix} = \begin{bmatrix} 0 & 1 \\ -a_1 & -a_2 \end{bmatrix} \begin{bmatrix} x_{e1} \\ x_{e2} \end{bmatrix} + \begin{bmatrix} 0 \\ b \end{bmatrix} u + \begin{bmatrix} 0 \\ N \end{bmatrix} \tag{1}$$

where $a_1 = 1/LC$, $a_2 = 1/RC$, $b = K_{PWM}/LC$, K_{PWM} is the inverter equivalent gain, and $N = -a_1 v_d - a_2 \dot{v}_d - \ddot{v}_d$ signifies the disturbance, and u denotes the control signal that can enforce the error to the origin within finite time.

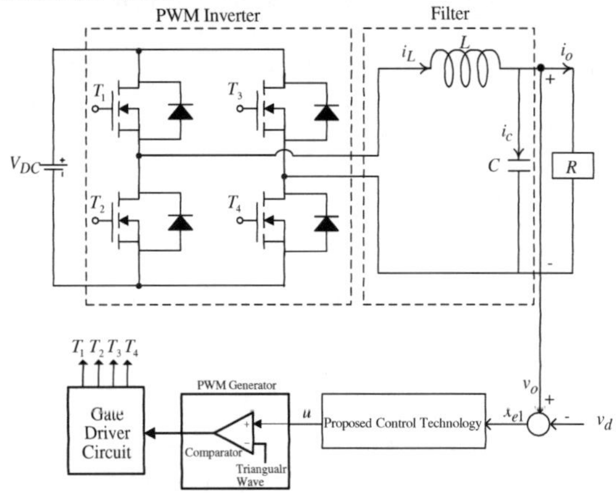

Figure 1. AC power source of PMM.

To speed up the convergence speed, a finite-time convergent sliding function $s_1(t)$ is defined due to the infinite system-state convergence time of the conventional sliding mode function $s(t) = c_1 x_{e1} + x_{e2}$ (c_1 is constant).

$$s_1(t) = x_{e1}(t) + \frac{1}{\mu} x_{e2}^{p_1/q_1}(t) \tag{2}$$

where $p_1 > q_1$, p_1 and q_1 stand for positive odd numbers ($1 < p_1/q_1 < 2$), and a sliding mode reaching term $\dot{s}_1 = -\sigma sign(s_1)$ is constructed.

Then, the control law u can be expressed as

$$u(t) = u_{eq}(t) + u_s(t) \tag{3}$$

with

$$u_{eq}(t) = b^{-1}[a_1 x_{e1} + a_2 x_{e2} - \mu \cdot \frac{q_1}{p_1} x_{e2}^{2-p_1/q_1}] \tag{4}$$

$$u_s(t) = -b^{-1}[\sigma sign(s_1)],\ \sigma > 0 \tag{5}$$

where u_{eq} is the equivalent control, and u_s displays the sliding control for compensating the perturbation influences. In practice, a smooth *approximation* of the *sign* function in (5) can be replaced by $\tanh(s_1) = \exp(s_1) - \exp(-s_1)/\exp(s_1) + \exp(-s_1)$.

Note that the chatter around the FTCSF still exists if a highly nonlinear loading is applied. Thus, to remove the chatter, the FOA is used to adjust the control gains of FTCSF. The rules of the FOA are supposed below: (1) The FOA is independent of sex; the all fireflies are unisex and one firefly can seduce other fireflies. (2) The attraction is proportioned to the firefly shine. The less bright firefly moves towards brighter firefly. If the distance is extended among fireflies, the brightness is lessened. The brightest firefly will move randomly since there are no insects to seduce it. (3) The brightness of the firefly correlates closely with the analytical form of the objective function. The brightness can proportion to the value of the objective function in a maximization problem.

While firefly j is more glamorous than firefly i, the activity is decided as

$$x_i = x_i + \phi \cdot e^{-\rho r_{ij}^2} \cdot (x_j - x_i) + \varepsilon \cdot (rand - 1/2) \tag{6}$$

where ϕ symbols the firefly attraction value, ρ represents the light absorption coefficient, $\varepsilon \in [0,1]$, and $rand$ signifies random number between 0 and 1.

The distance r_{ij} between fireflies i and j can be written by using Cartesian distance below.

$$r_{ij} = \left\| x_i - x_j \right\| \tag{7}$$

Thus, the FOA compares the attraction of the firefly i with the attraction of the firefly j. If the firefly j is more glamorous than firefly i, then the firefly i is moved to the new place; otherwise the firefly i will stay in the present place.

3. Experimental results

The proposed system parameters are given in Table 1. Under nonlinear load (full wave rectified with a parallel resistor and capacitor), the output waveforms with the proposed control technology and the conventional sliding mode are shown in Figure 2 and Figure 3, respectively. The output voltage of the proposed system has lower distortion than that of the conventional sliding mode controlled system. The good steady-state and dynamic response can be obtained by using the proposed control technology; the output voltage close to sine wave achieves a low %THD of 1.43%, with the proposed control technology while the output voltage with the conventional sliding mode yields a high %THD of 9.02%. Because the chatter is removed remarkably, Figure 4 shows that the tracking errors with the proposed control technology can be fast converged to the origin, however there is a great deal of tracking errors in the conventional sliding mode controlled system.

Table 1. System parameters

Filter inductor	L=0.1 mH
Filter capacitor	C=10 μF
DC supply voltage	V_{DC}=200 V
Nonlinear load	Electrolytic Capacitor: 120 μF; Load Resistor: 35 Ω
Output voltage and frequency	v_o=110 V_{rms}, f=60 Hz
Switching frequency	f_{sw} =18 kHz

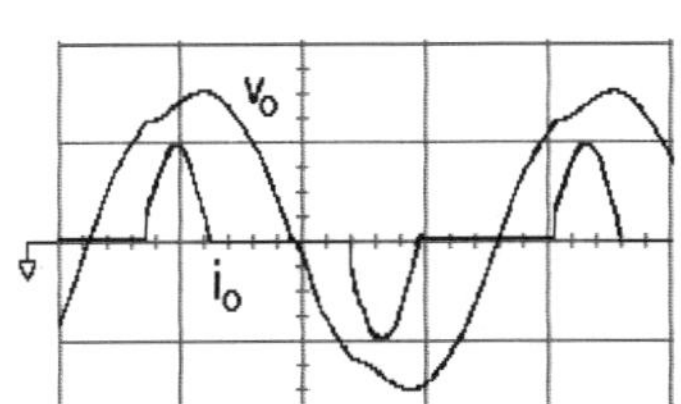

Figure 2. Proposed control technology with nonlinear load (100V/div; 25A/div; 5ms/div)

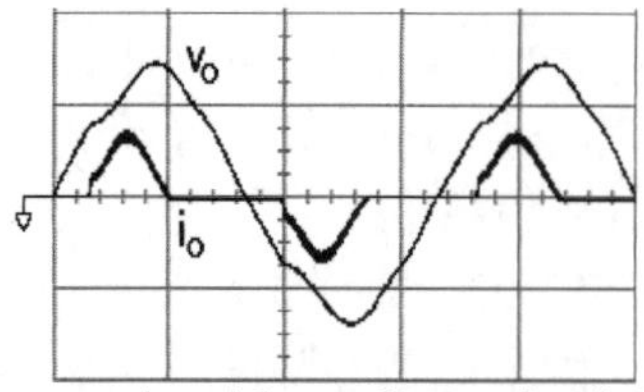

Figure 3. Conventional sliding mode with nonlinear load (100V/div; 25A/div; 5ms/div)

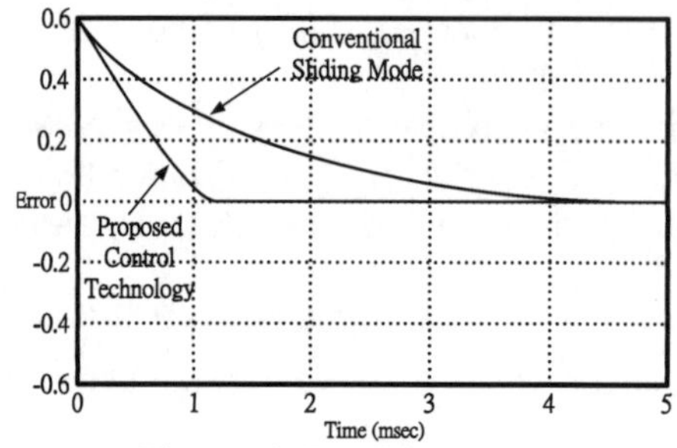

Figure 4. Tracking errors

4. Conclusions

In this paper, a FTCSF with FOA is presented for the application of AC power source of PMM. Relative to the conventional sliding mode, the proposed control technology can remove the chatter, increase the convergence speed, achieve satisfactory steady-state, and fast transience. Experimental results reveal that low THD, fast transient response, the reduction of the chatter, and the attenuation of the tracking error are obtained in the proposed controlled system even under highly nonlinear loading.

5. Acknowledgment

This work was supported by the Ministry of Science and Technology of Taiwan, R.O.C., under contract number MOST106-2221-E-214-031.

6. References

[1] Branko L D and Branko B 2015 *Power Electronics: Converters and Regulators* (New York: Springer)

[2] Shin H B and Park J G 2012 Anti-Windup PID Controller with Integral State Predictor for Variable-Speed Motor Drives *IEEE Trans. Industrial Electronics* vol. 59, no. 3 pp 1509-1516

[3] Saleh S A and Rahman M A 2011 Experimental Performances of the Single-Phase Wavelet-Modulated Inverter *IEEE Trans. Power Electronics* vol. 26, no. 9 pp 2650-2661

[4] Hu J B and Zhu Z Q 2013 Improved Voltage-Vector Sequences on Dead-Beat Predictive Direct Power Control of Reversible Three-Phase Grid-Connected Voltage-Source Converters *IEEE Trans. Power Electronics* vol. 28, no. 1 pp 254-267

[5] Zhang B Zhou K L and Wang D W 2014 Multirate Repetitive Control for PWM DC/AC Converters *IEEE Trans. Industrial Electronics* vol. 61, no. 6 pp 2883-2890

[6] Azar A T and Zhu Q M 2015 *Advances and Applications in Sliding Mode Control Systems* (New York: Springer)

[7] Yu X H and Xu J X 2002 *Variable Structure Systems: Towards the 21st Century* (New York: Springer)

[8] Galias Z and Yu X H 2012 Dynamical Behaviors of Discretized Second-Order Terminal Sliding-Mode Control Systems *IEEE Trans. Circuits and Systems II: Express Briefs* vol. 59, no. 9 pp 597-601

[9] Tamilselvan R and Padmathilagam V 2014 Harmonic Optimization in Cascaded Multi Level Inverter Using Firefly Algorithm *Journal of Theoretical and Applied Information Technology* vol. 64, no. 3 pp 773-780

[10] Yang X S 2008 *Nature-Inspired Metaheuristic Algorithms* (Frome: Luniver Press)

IOP Conf. Series: Materials Science and Engineering **311** (2018) 012006 doi:10.1088/1757-899X/311/1/012006

Kinetic analysis of elastomeric lag damper for helicopter rotors

Yafang Liu[1], Jidong Wang[1] and Yan Tong[1]
[1] Beihang University, Xueyuan Road No.37, Beijing 100191, China

Abstract. The elastomeric lag dampers suppress the ground resonance and air resonance that play a significant role in the stability of the helicopter. In this paper, elastomeric lag damper which is made from silicone rubber is built. And a series of experiments are conducted on this elastomeric lag damper. The stress-strain curves of elastomeric lag dampers employed shear forces at different frequency are obtained. And a finite element model is established based on Burgers model. The result of simulation and tests shows that the simple, linear model will yield good predictions of damper energy dissipation and it is adequate for predicting the stress-strain hysteresis loop within the operating frequency and a small-amplitude oscillation.

1. Introduction

Elastomeric lag dampers are vital components of helicopter. They can suppress the ground resonance and air resonance that play an significant role in the stability of the aircraft. They are used in well-known aircraft such as McDonnell AH-64 Apache, Bell Model 412 and Boeing Model 360 [1]. Due to the widespread employment of the lag damper as critical element of the rotor, the modeling of the elastomeric lag dampers have received increasing attention in recent years.Hu,Xiang and Zhang [2] analyzed elastomeric lag damper kinematic coupling and its effect on helicopter air resonance in hover. Chellil et.al [3] compared elastomeric damper to other hydraulicdamper,and they presented that elastomeric damper produced better results to reduce the vibration level in the helicopter rotors. Byers and Gandhi [4] believed that the radial absorbers is more effective in transferring damping to the rotor blade lag mode than chordwise absorbers.

For many years, the complex modulus method has been applied in describing the stiffness and damping properties of elastomers. The method builds the model of elastomeric lag damper by operating a linear spring and linear damper in parallel. However, we can know that the elastomeric is a kind of nonlinear material through the relevant researches of elastomeric properties.An in-depth discussion that the nonliear charateristic of elastomer have an effect on design of lag damper is presented in article writted by Hausmann [6].

In order to solve this problem, some researchers have put forward some solutions. Felker [5] propose the expression of elastomeric lag damper force by the superposition of spring force which is expressed by a nonlinear function of displacement and damper force which is expressed by a nonlinear function of displacement and velocity. Others make a break with the complex modulus approach. Gandhi and Chopra [7] place an additional nonlinear spring in series with the linear, parallel spring and dashpot. The method for representing the series spring by a quadratic function of displacement is purely to improve the precision of the static response. However, it does not account for nonlinear damping. Other researchers [8] also used some methods that different with the complex modulus approach to simulate the elastomeric materials. However, they almost have got the same model as Gandhi and Chopra proposed. This method is based on the elastic displacement fields, which is extended to model strain-dependent material behavior. The mechanical analogy of the model [9] is structurally the same as the model of Gandhi and Chopra, except that both springs and dampers are now represented by nonlinear functions. These methods are only available to solve some problems aimed to specific design parameters, rather than a comprehensive and common damper model.

In this paper, we establish the dynamic model in time-domain by finite element to obtain a more common model and evaluate their performance by comparing the analysis with the test measurements of transient blade motions. Firstly, we analyze stress-strain relationship when shear force is subjected to elastomeric lag damper which is made from silicone rubber. Within the operating frequency and a small displacement amplitude range, stress-strain hysteresis loop of elastomeric lag damper, which is based on ABAQUS, can be simulated when harmonic displacement is excited to silicone rubber.

2. The time-domain modeling of Elastomeric Damper

In this paper, we study the stress-strain hysteresis curve of elastomeric material with small amplitude applied shear force at a certain frequency. In order to simplify the model more conveniently, we can use the linear modeling of the elastomeric material to approximate the real material. The simplest elastomeric model is a spring and a damper in series or in parallel, named Maxwell model and Kelvin model. Maxwell model can express the relaxation elastomeric material, but cannot represent the shadow of creep. It reflects the steady flow only. In contrast, Kelvin model can express the creep, but cannot reflect the stress relaxation.

2.1. Burgers model

In this paper, we used a four-parameter model with the Maxwell and Kelvin unit in series, named Burgers model. Although there are some big differences between this model and the real elastomeric material, it can represent two main characteristics of silicone rubber. Figure 1 is a Burgers model.

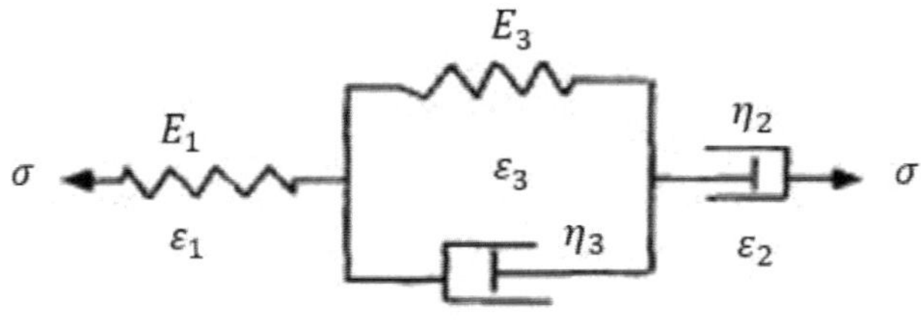

Figure 1. Burgers model

The deformation of Burgers model includes elastic deformation, viscous deformation and elastomeric deformation, the stress-strain relationship:

$$\varepsilon_1 = \sigma / E_1 \tag{1a}$$

$$\sigma = \eta_2 \dot{\varepsilon}_2 \tag{1b}$$

$$\sigma = E_3 \varepsilon_3 + \eta_3 \dot{\varepsilon}_3 \tag{1c}$$

Total strain:
$$\varepsilon = \varepsilon_1 + \varepsilon_2 + \varepsilon_3 \tag{2}$$

Using differential operator method, the constitutive equation of Burgers model is transformed into:

$$\sigma + p_1\dot{\sigma} + p_2\ddot{\sigma} = q_1\dot{\varepsilon} + q_2\ddot{\varepsilon} \tag{3}$$

In this equation,

$$p_1 = \frac{\eta_2}{E_1} + \frac{\eta_2 + \eta_3}{E_3};\ p_2 = \frac{\eta_2\eta_3}{E_1 E_3};\ q_1 = \eta_2;\ q_2 = \frac{\eta_2\eta_3}{E_3}.$$

2.2. The parameters of Burgers model

IOP Conf. Series: Materials Science and Engineering **311** (2018) 012006 doi:10.1088/1757-899X/311/1/012006

Using Laplace transform with formula (3), we can infer the equation of the creep compliance in Burgers model :

$$J(t)=\frac{1}{E_1}+\frac{t}{\eta_2}+\frac{1}{E_3}(1-e^{-\frac{E_3 t}{\eta_3}}) \tag{4}$$

With formula (4), if creep compliance value is known each time, the least squares method can be used to fit out parameters E_1,η_2,E_3,η_3 of elastomeric material. Specific steps: we choose the initial value of elastic modulus and viscosity coefficient. Furthermore, the approach that solute iteratively with MATLAB and optimize the value of four parameters can be used.

2.3. Transform the parameters of Burgers model into Prony series

Because the parameters from the fitting model cannot be read into finite element software directly. We want to use the software to simulate the mechanical characteristics of elastomeric material, we need transform the parameters of Burgers model into Prony series.

Supposing the volume strain is thoroughly recoverable and elastic deformation, so the normalized bulk modulus ki in Prony series are 0. Supposing shear strain was rheological, we can obtain:

$$\begin{aligned} S_{ij} &= Y'(t)*de_{ij}=2G(t)e_{ij} \\ \sigma_{ii} &= Y''(t)*de_p=3K\varepsilon_{ii} \end{aligned} \tag{5}$$

or

$$\begin{aligned} e_{ij} &= J'(t)*dS_{ij} \\ \varepsilon_{ii} &= J''(t)*d\sigma_{ii} \end{aligned} \tag{6}$$

S_{ij} and e_{ij} are the deviatoric tensor of stress and the deviatoric tensor strain. σ_0 , e_0 are the spherical tensor of stress and the spherical tensor of strain,* represents Stieltjes convolution.

In Burgers model:

$$G(t)=\frac{Y'(t)}{2}=\frac{G_1}{\alpha-\beta}[(\frac{G_2}{\eta_3}-\beta)e^{-\beta t}-(\frac{G_2}{\eta_3}-\alpha)e^{-\alpha t}] \tag{7}$$

$$J'(t)=\frac{1}{2G_1}+\frac{t}{2\eta_2}+\frac{1}{2G_2}(1-e^{-\frac{G_2 t}{\eta_3}}) \tag{8}$$

where

$$\alpha=\frac{p_1+\sqrt{p_1^2-4p_2}}{2p_2}$$

$$\beta=\frac{p_1-\sqrt{p_1^2-4p_2}}{2p_2}$$

According to the generalized Hooke's law and the corresponding expression of elastomeric material , the Laplace solution with uniaxial compression can be obtained:

$$\bar{\varepsilon}_{xx}=\frac{1}{E^*}\sigma_{xx}=\frac{s[2\bar{J}'(s)+\bar{J}''(s)]}{3}\frac{\sigma_0}{s}=[\frac{2}{3}\bar{J}'(s)+\frac{1}{3}\bar{J}''(s)]\sigma_0 \tag{9}$$

Substituting equation(6), equation(8)and $K=\frac{2(1+\mu)}{3(1-2\mu)}G_1$ into equation (9):

$$J^*=\frac{2J'(t)+J''(t)}{3}=\frac{1}{2(1+\mu)G_1}+\frac{t}{3\eta_2}+\frac{1}{3G_2}(1-e^{-\frac{G_2 t}{\eta_3}}) \tag{10}$$

In order to distinguish with one-dimensional elastomeric parameters, we replace η with n . By comparing to equation(4), we can infer:

$$G_1 = \frac{E_1}{2(1+\mu)} \ , \ n_2 = \frac{\eta_2}{3} \ , G_2 = \frac{E_3}{3} \ , \ n_3 = \frac{\eta_3}{3} \tag{11}$$

Substituting equation（7）, the shear modulus of Burgers model can be transformed into Prony series：

$$G(t) = G_\infty + G_1 e^{-\frac{t}{\tau_1}} + G_2 e^{-\frac{t}{\tau_2}} \tag{12}$$

Where $G_\infty = G_0 - G_1 - G_2$, $\tau_1 = 1/\alpha$, $\tau_2 = 1/\beta$,and τ_1, τ_2 reflect the tao1and tao2 in ABAQUS. In ABAQUS, $g1 = G_1 / G_0$, $g2 = G_2 / G_0$, the values of k1and k2 are 0.

3. Finite element modeling of dynamics of elastomeric damper

This paper mainly focuses on the analysis and calculation of plate-formed elastomeric damper. The simplified model is as follows:

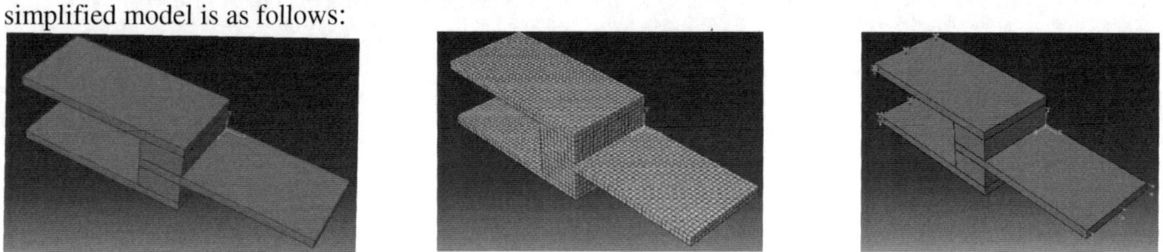

(a)Model diagram (b) Grid diagram (c)Boundary condition diagram

Figure 2. Simplified model of plate elastomeric lag damper

The upper、lower and the intermediate plate are all made by aluminum alloy, and filled by silicone rubber material. Each piece of aluminum alloy and silicone rubber is a separate component, and the five component combine into a whole entity which is spliced by tie-binding command.

At first, we should define the materials parameters. A series of unified unit N-mm-MPa are employed. And the parameters are listed as follows:

Table 1 basic materials parameters.

Materials	Aluminum alloy plate	Silicone rubber
Modulus of elasticity	70Gpa	4MPA
Poisson's ratio	2.7	0.475

Though calculation, hyperelastic parameters are settled as: $C_{10} = 0.176, C_{01} = 0.00432, D_1 = 0$. And the viscoelastic creep parameters are from Bin Zhang. By fitting on creep compliance in MATLAB, we can obtain the Burgers model parameters :

$E_1 = 0.557MPa, \eta_2 = 291\times10^3 MPa/s, E_3 = 0.438MPa, \eta_3 = 5.94\times10^3 MPa/s$.

Using the method as mentioned in the last part, we can transform the parameters of Burgers model into Prony series :

$g1 = G_1 / G_0$=0.1525, $g2 = G_2 / G_0$=0.2567, k1=k2=0, $\tau_1 = 1/\alpha$=0.0009507. $\tau_2 = 1/\beta$=6.0332.

We need to set the boundary conditions for the model. The upper and lower aluminum alloy plates are set the whole constraint at the left end to fix the two plates. Since the shear stress-strain hysteresis curves are dependent on the displacement, we apply different frequency(3Hz,6Hz,9Hz) to the intermediate aluminum plate. The curves of stress and strain changed with time can be obtained. Then we export the data to EXCEL to draw the force-displacement curves.

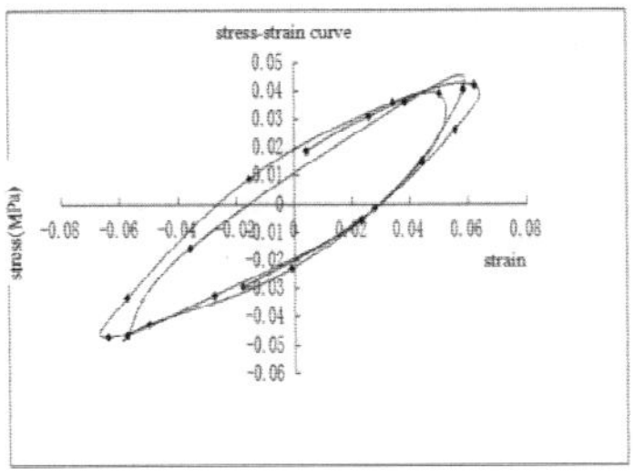

Figure 3. 3HzStress-Strain curve

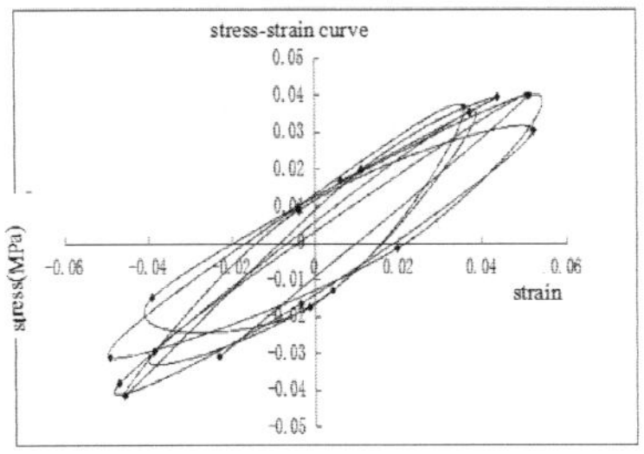

Figure 4. 6Hz Stress-Strain curve

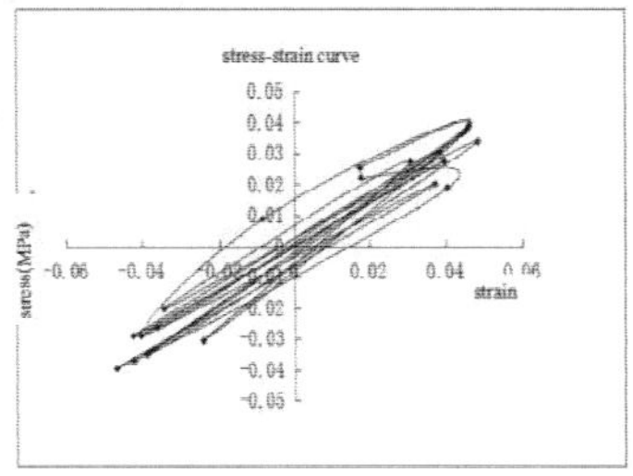

Figure 5. 9Hz Stress-Strain curve

4. Elastomeric Lag Damper dynamic response tests

4.1. Experimental subject

(1) Verify ther force-displacement hysteretic curves of elastomeric lag damper;(2) Discover the variation discipline of elastomeric lag damper parameters

4.2. Geometric parameters of Elastomeric Lag Damper plate

The silicone rubber and metal plate are spliced by vulcanization at room temperature. It can be used when it is cured after 36 hours.

Table 2. Geometric parameter of plate elastomeric lag damper model.

Component parameters	thickness (mm)	length (mm)	width (mm)	elasticity modulus (MPa)	Poisson 's ratio
Intermediate aluminium plate	6	56	41	70000	0.3
Upper and lower aluminum plates	4	41	41	70000	0.3
silicon rubber	8	33	41	4	0.475

Experimental equipment diagram and photographs are showned in Figure 6 and Figure 7.

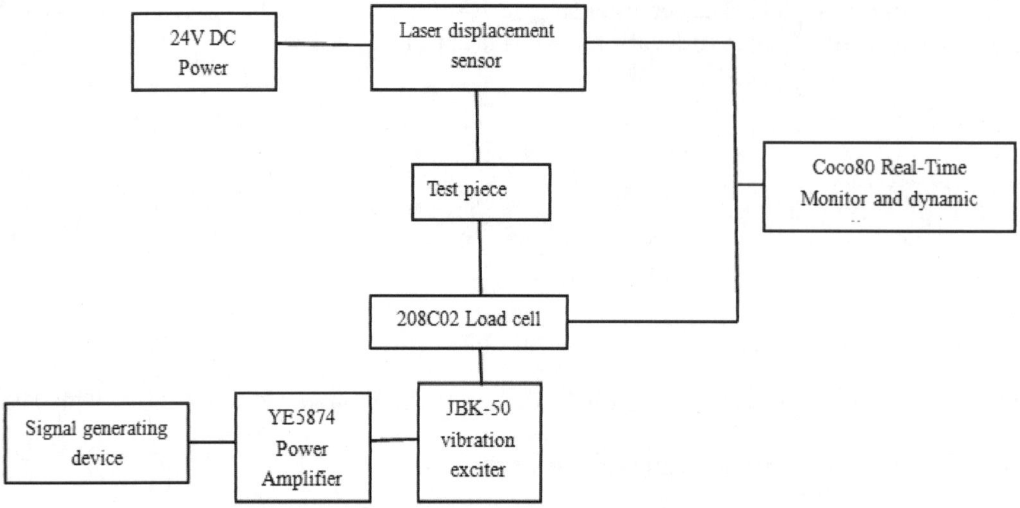

Figure 6. experimental equipment diagram

(a) (b)

Figure 7. Experimental photographs

4.3. Experiment methods

Adopting the method of sinusoidal excitation, the experiment is loaded with displacement as the controlled quantity. By applying different frequency sinusoidal force, we measure the shear displacement and force of the elastomeric lag damper under the same shear strain amplitude, respectively. Therefore, we can obtain the resulting of laws of lag dampers with different frequency. During the experiments, force sensor and laser displacement sensors was used to measure and Coco80 dynamic data acquisition and real-time monitoring system is used to record the force and displacement. Prediction of shear deformation U is about 0.4 mm. Vibration frequency is 3Hz, 6Hz and 9Hz, respectively. Sampling frequency is 2.56 KHz, sampling points is 2560.

The experiments steps are as follow:

(1) Install the sample and ensure the specimens are installed correctly;

(2) Install the force sensor and ensure that the surface of the force sensor was attached closely to the

sample to make sure the measuring accuracy. Then contacted the displacement sensor and Coco80 and use the channels 2.

(3) Install the laser displacement sensor and contact it with the 24V DC power and the YE5874 power amplifier. Then contact the YE5874 power amplifier and Coco80, using the channels 1.
(4) Contact the JBK-50 vibration exciter and YE5874 power amplifier.
(5) Contact Coco80 and YE5874 power amplifier.
(6) Set the Coco80 frequency to 3HZ. Adjust the shear displacement input and control the amplitude at around 0.8mm. Then start to apply a sinusoidal force and use laser displacement sensor and force sensor to measure the shear deformation and damping force. In the meantime, using data collection devices to record the data of force and shear deformation.
(7) Keep the shear displacement constant. Transform the Coco80 input frequency into 6HZ and 9HZ, respectively. Then repeat the experiment from step 2.

4.4. Experiment results

From the EXCEL analysis of collected data, we can find that the force-displacement curves (Figure 8, 9 and 10) are nearly an oval in all the cases, which shows that the specimens have a favorable energy-absorbing effect.

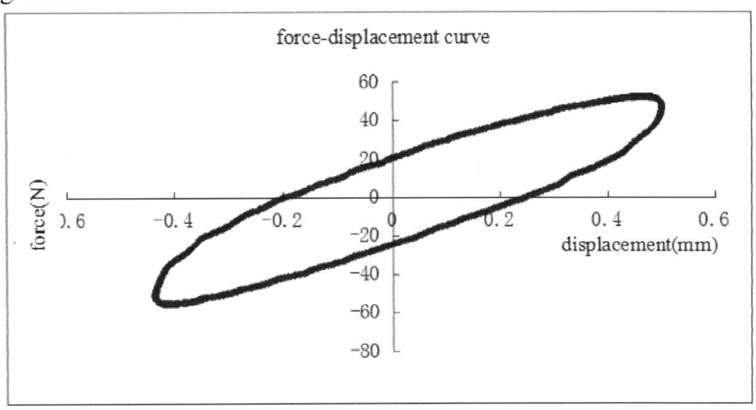

Figure 8. 3Hz Force-Displacement curve

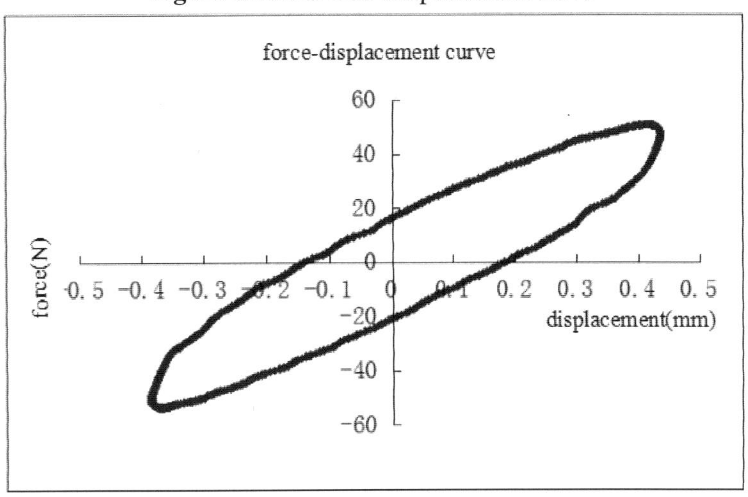

Figure 9. 6Hz Force-Displacement curve

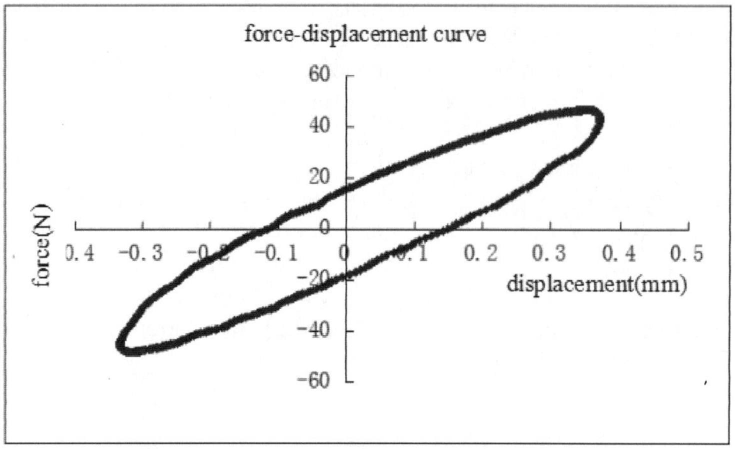

Figure 10. 9Hz Force-Displacement curve

The force-displacement curves show that the slope of the ellipse increases with the increasing of the vibration frequency when the temperature and shear displacement amplitude is a constant. Namely, the stiffness of the elastomeric lag damper model increases with the increasing of the vibration frequency.

4.5 Analysis

From the experiment we can get the conclusion as follow:

(1) Dynamic characteristics of the elastomeric lag damper will change with the vibration frequency and the amplitude of the shear displacement.

(2) When the elastomeric lag damper is subjected to alternating stress, the strain phase is lagging behind the stress phase from 0 and $\pi/2$

(3) Under the small shear displacement amplitude, elastomeric lag damper stress-strain hysteresis curve is an ellipse.

(4) As the frequency increased, the area of the stress-strain curves decreased. Namely, the energy consumption decreased.

5. Comparison

We can transform the force-displacement curve drawn by experimental results into stress-strain curve by the formula as follow:

$$\sigma = \frac{F}{A} \qquad \varepsilon = \frac{u}{t} \tag{13}$$

where F and u are resilience and shear displacement measured by experiments, respectively; and A that is shear area is equal to $2\times 1353\,mm^2$, and t that is the thickness of the silicone rubber is equal to 8.

By comparing the results of simulation to the results of experiment, we can draw three pictures as follows:

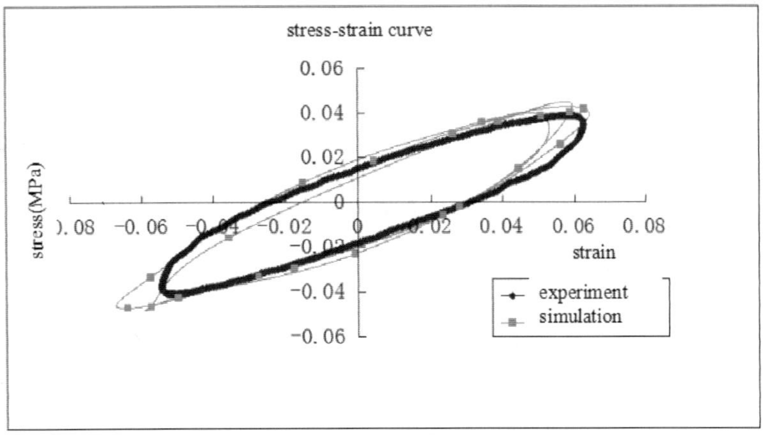

Figure 11. 3Hz Comparison of ABAQUS and experimental results

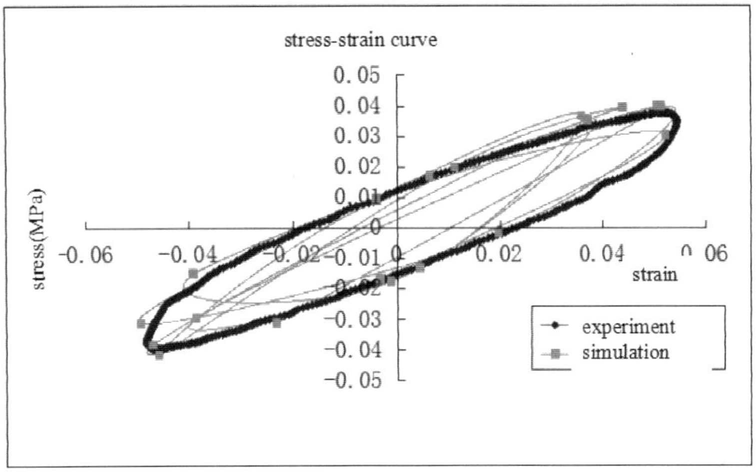

Figure 12. 6Hz Comparison of ABAQUS and experimental results

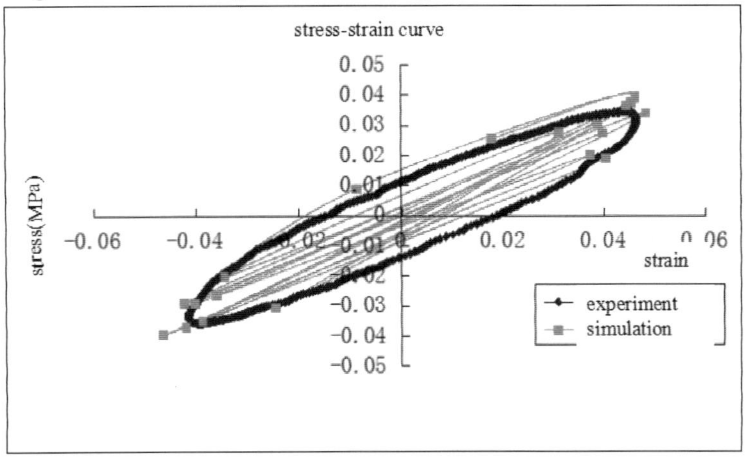

Figure 13. 9Hz Comparison of ABAQUS and experimental results

Based on comparisons of the results of the experiments and the dynamics simulations of the various lag damper models, the following conclusions can be made:

1) The two hysteresis loops curves are coincided obtained by the experiment and ABAQUS

IOP Conf. Series: Materials Science and Engineering **311** (2018) 012006 doi:10.1088/1757-899X/311/1/012006

simulation .It is verified that the model used in the thesis is reliable.
2) Within the low frequency and a small displacement amplitude range,the stiffness is increasing while the frequency is amplifying.

6. Conclusions

In this study, well-known techniques, for the standard experimental modal analysis and correlation of finite element models with test data, combined with the Burgers method have been used to identify the time-dependent properties of a elastomeric material. The linear modeling approach that employed Burgers Models to evaluate the behavior of elastomeric dampers will produce accurate estimates of energy dissipation, but it is not sufficient for prediction of large deformation response.
A more in-depth investigation of some aspects, important for actual applications, was postponed to future work, including a discussion of the effects of nonlinear properties of elastomeric lag dampers when they experienced a high frequency and a large deformation.

7. References

[1] McGuire D P 1976 The Application of Elastomeric Lead-Lag Dampers to Helicopter Rotors *Lord Corporation Library of Technical Papers* (Erie, Pennsylvania)
[2] Hu G C, Xiang J w, Zhang X G 2002 Analytical Model of Elastomeric lag Damper kinematic coupling and its effect on helicopter air resonance in hover *Chinese J. Aeronautics* **15** 27-32
[3] Chellil A, Nour A, Lecheb S, Kebir H and Chevalier Y 2013 Impact of the fuselage damping characteristics and the blade rigidity on the stability of helicopter *Aerospace Sci. Technol* **29** 235-52
[4] Byners Land Gandhi F 2009 Embedded absorbers for helicopter rotor lag damping *J. Sound and Vibration* **325** 705-21
[5] Felker F F, Lau B H, McLaughlin S, *et al. 1987* Nonlinear behavior of an elastomeric lag damper undergoing dual-frequency motion and its effect on rotor dynamics *J. AHS.***32** 45-53
[6] Hausmann G 1986 Structure analysis and design considerations of elastomeric dampers with viscoelastic material behavior *Proc. 12th. European Rotorcraft Forum* (Federal Republic of Germany :Garmich-Partenirchen)
[7] Gandhi F and Chopra I 1994 Elastomeric lag damper effects fag-lag stability in forward flight. *Proc. 35th AIAA/ASME/ASCE/AHS/ASC Structures* (Hilton Head: Structural Dynamics and Materials Conference) p 2726-38
[8] Tarzanin F J and Panda B 1995 Development and application of nonlinear elastomeric and hydraulic lag damper models. *Proc. 36th AIAA/ASME/ASCE/AHS/ASC Structures*(New Orleans: Structural Dynamics and Materials Conference) p 2534-44
[9] Kunz D L 1997 Influence of elastomeric damper modeling on the dynamic response of helicopter rotors *AIAA Journal* **35 349-54**

IOP Conf. Series: Materials Science and Engineering **311** (2018) 012007 doi:10.1088/1757-899X/311/1/012007

Optimal design of geometrically nonlinear shells of revolution with using the mixed finite element method

L U Stupishin , K E Nikitin and A G Kolesnikov
Urban, road construction and structural mechanic department, South-West State University (SWSU), 305040, Kursk, Russia

Abstract. The article is concerned with a methodology of optimal design of geometrically nonlinear (flexible) shells of revolution of minimum weight with strength, stability and strain constraints. The problem of optimal design with constraints is reduced to the problem of unconstrained minimization using the penalty functions method. Stress-strain state of shell is determined within the geometrically nonlinear deformation theory. A special feature of the methodology is the use of a mixed finite-element formulation based on the Galerkin method. Test problems for determining the optimal form and thickness distribution of a shell of minimum weight are considered. The validity of the results obtained using the developed methodology is analyzed, and the efficiency of various optimization algorithms is compared to solve the set problem. The developed methodology has demonstrated the possibility and accuracy of finding the optimal solution.

1. Introduction

The proposed methodology of optimal design of shells allows designing the shells with variable thickness, e.g., ribbed shells, shells with a discrete inner layer, etc. In this paper the shell is regarded as structurally orthotropic.

Stress-strain state of shell for the solution of optimization problem was determined using the Galerkin method [1] in mixed finite element formulation [2, 3]. This method allows to do without the construction procedure of the functional of problem and to solve equations of problem in the form in which they are written. The authors were obtained in an explicit form of the matrix and vectors for the finite element shell. Their advantage is that the numerical integration is not used, which improves the accuracy of calculations. Another of the advantages of using the mixed formulation (as indicated by, e.g., Streng [2]) - reduction of the condition number of the matrix of the finite element, in comparison with the classical method of finite elements in displacements, which also positively effect on the accuracy of the calculation results. The methodology for determining the stress-strain state of geometrically nonlinear shells of revolution is described in detail in [3-5].

2. Description of the methodology

The problem of optimal design of the shell was considered as the problem of shell volume minimization with strength, stability and displacement constraints:

$$\bar{V}(\bar{x}) = 2\pi \sum_{i=1}^{N} \int_{-1}^{1} h_i(\rho)\rho \left(1 + \left(\frac{\mathrm{d}f(\rho)}{\mathrm{d}\rho}\right)^2\right)^{1/2} \mathrm{d}\xi \to \min ,$$

$$p - p_{\mathrm{cr}}(\bar{x}) \geq 0 ,$$

$$w_{max}(\bar{x}) - w_u \le 0\,,$$

$$\sigma_{max}(\bar{x}) - R \le 0\,. \tag{1}$$

Such parameters as shell shape parameters, thickness distribution, etc. can be selected as components of the vector of the design variables $\bar{x}$.

In equation (1) p is the load imposed on the shell; $p_{cr}(\bar{x})$ is the critical load; $w_{max}(\bar{x})$ is the maximum deflection of the shell; w_u is the ultimate permissible deflection; $\sigma_{max}(\bar{x})$ is the maximum equivalent stresses in the shell; R is the ultimate allowable stress.

The component of the vector of maximum equivalent stresses ($\sigma_{max}(\bar{x})$) are calculated as:

$$\sigma_{max} = \left[\left(\frac{N_\alpha}{h(\alpha,\beta)} + \frac{6M_\alpha}{h(\alpha,\beta)^2} \right)^2 + \left(\frac{N_\beta}{h(\alpha,\beta)} + \frac{6M_\beta}{h(\alpha,\beta)^2} \right)^2 - \right.$$

$$\left. - \left(\frac{N_\alpha}{h(\alpha,\beta)} + \frac{6M_\alpha}{h(\alpha,\beta)^2} \right) \cdot \left(\frac{N_\beta}{h(\alpha,\beta)} + \frac{6M_\beta}{h(\alpha,\beta)^2} \right) + 3 \cdot \left(\frac{S}{h(\alpha,\beta)} + \frac{6H}{h(\alpha,\beta)^2} \right)^2 \right]^{1/2} \tag{2}$$

The values $p_{cr}(\bar{x})$, $w_{max}(\bar{x})$ and $\sigma_{max}(\bar{x})$ are determined at each step of the optimization algorithm using the Galerkin method in mixed finite element formulation [4 - 6].

The problem of optimal design with constraints is reduced to the problem of unconstrained minimization using the penalty functions method. Penalty function r_k takes the form proposed by Mischke [6]:

$$\bar{V}_k(\bar{x}) = V(\bar{x}) + r_k d \sum_{i=1}^{N} b_i \cdot (\bar{x}_c - \bar{x})^2 + s \to \min, \tag{3}$$

where, $V(\bar{x})$ is the objective function of the problem with constraints; $\bar{x}_c$ is the vector of design parameters values corresponding to any point in design space, where all constraints $g_i(\bar{x})$ hold (in the domain of feasible solutions); N is the number of constraints; d is the scale multiplier to reduce functions $V(\bar{x})$ and $g_i(\bar{x})$ to same order; b_i is the parameter equal to 1 if the constraint $g_i(\bar{x})$ is satisfied and equal to 0 if it is not satisfied; r_k is the penalty parameter for the k-th minimization problem (iteration) selected from the following sequence: 1, 100, 10,000, 1,000,000; s is the parameter used in some problems to define a barrier on the boundary of feasible solutions (s=0 within the domain of feasible solutions and s=const>0 outside this domain); k=1..M – number of iteration; M is the number of iterations.

The problem of optimal design in form as in equation (3) can be solved any of classical algorithms of unconstrained minimization (e.g. method of steepest descent, grid search method, method of conjugate gradients, etc.).

3. Test problems

To check the validity of the results obtained using the developed methodology and to compare the effectiveness of different algorithms of unconstrained optimization, two test problems of optimal design of a shallow axisymmetric shell of revolution were solved. Shell dimensions: the shell base radius a=3 [m], rising height f_0=0.45 [m]; shell material characteristics: E=2.1·1011[Pa], ν=0.3, R=210 [MPa]; ultimate strain w_u=0.01 [m]. The shell is closed at the top. The shell is fixed rigidly to the support. The load is distributed uniformly on the shell; its intensity is p =10 [kPa]. The shell was divided evenly into 20 finite elements along the generatrix. The position the of the finite elements nodes along the shell generatrix was calculated by formula: $f(\rho) = f_0\rho^Z$, where Z is the shell shape parameter, ρ is the radial coordinate.

The 1st test problem is to determine the shape and thickness of the shell of minimum weight (design variables are Z and h) under the assumption that the shell thickness is the same over its entire surface.

The 2nd test problem is to determine the optimum thickness distribution for a minimal weight shell (design variables are h and $h1$) under the assumption that the thickness varies linearly from the centre (h1) to the edges (h) and does not change in the circumferential direction.

4. Checking the results

The diagrams of changes of objective and penalty functions depending on the node parameter values were plotted (see figure 1 - 5 for the 1st problem, and figure 6 for the 2nd problem).

For problem 1 the boundary of permissible values with stability, strain and strength constraints has the form of a parabola with the vertex at the values Z = 1.5 – 2 (figure 1 - 5). Taking into account the descending direction of the volume function, it is possible to determine visually the position of the objective function minimum on the boundary of the domain of permissible values when Z = 1 - 1.7 (see figure 5).

The results of the 1st optimization problem solution using the developed methodology are the values of design variables Z = 1.65, h = 0.00173 and the objective function value of 0.0075. The reliability of these results is confirmed by visual analysis (figure 5).

For 2nd problem the boundary of permissible values has the form close to a straight line (figure 6). Taking into account the descending direction of the volume function the minimum of the objective function is on the boundary of the search domain with the minimum values of $h1$ (figure 6).

The results of solving 2nd problem using the developed optimization methodology are the values of design variables $h1$ = 0.0005, h = 0.00756 when the value of the objective function is 1.07. This solution is consistent with a shell whose thickness decreases towards the centre. The validity of these results can be confirmed by the visual analysis of the diagram of the objective function changes (figure 6).

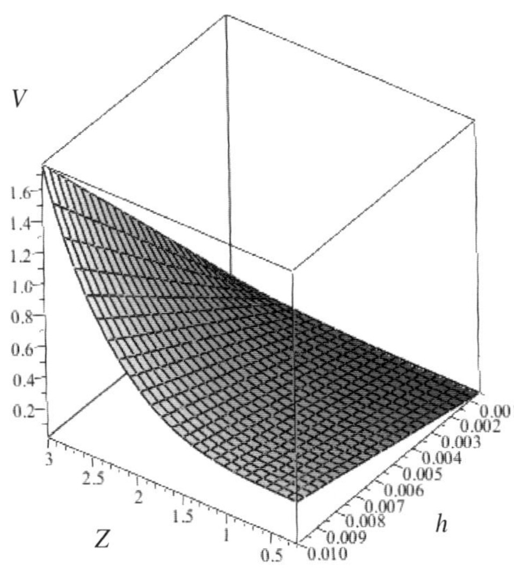

Figure 1. Dependence of shell volume change on parameters Z and h.

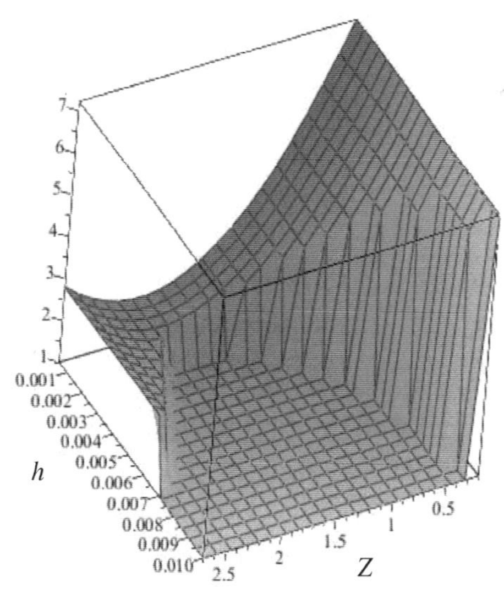

Figure 2. Dependence of penalty function for stability constraints on parameters Z and h.

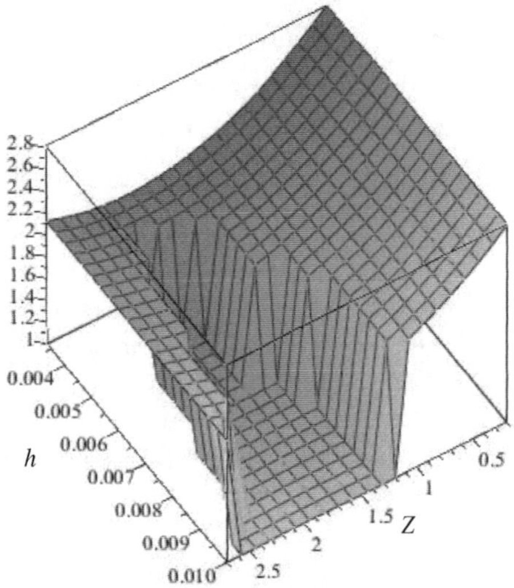

Figure 3. Dependence of penalty function for strength constraints on parameters Z и h.

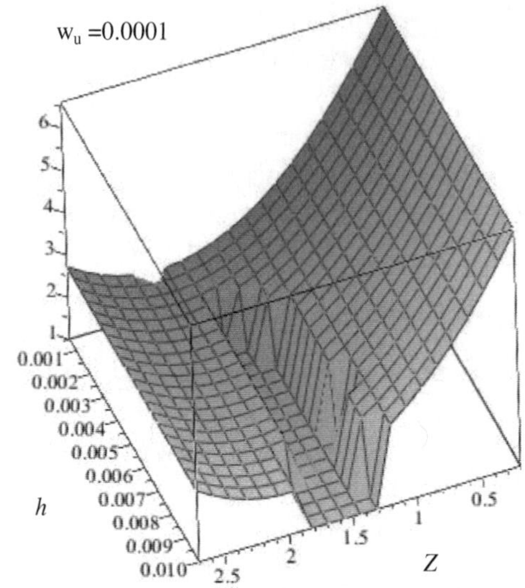

Figure 4. Dependence of penalty function for strain constraints on parameters Z и h.

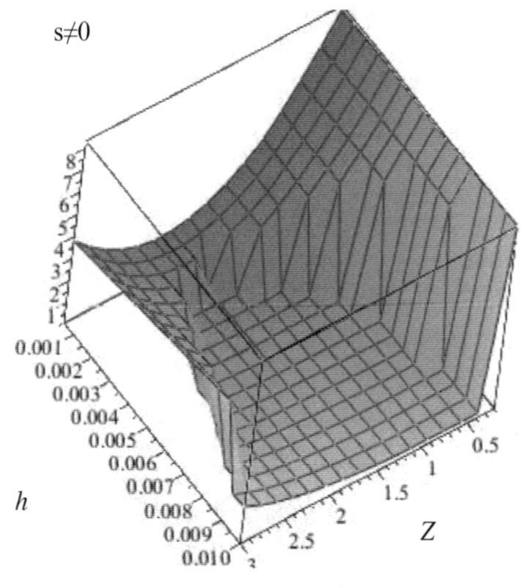

Figure 5. Dependence of 1st problem objective function on parameters *Z* and *h***.**

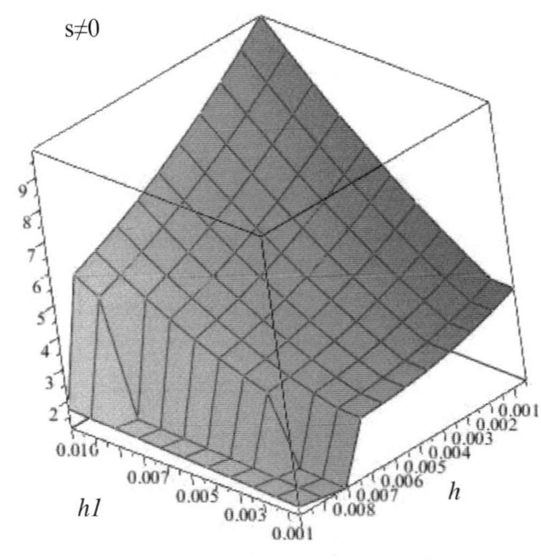

Figure 6. Dependence of 2nd problem objective function on parameters *h1* and *h***.**

When studying the shape of the boundary of the permissible values domain with regards to stability, minor irregularities associated with the inaccuracy of critical load determination algorithm were found. They can cause difficulties when determining an optimal solution. Therefore, to obtain reliable results search for an optimal solution should be started from several points, and then the best solution should be selected.

5. Analysis of efficiency of optimization algorithms

A comparative analysis of the effectiveness of different optimization algorithms used to solve fist problem have been carried out ([7-11]). Such methods as the grid (multidimensional continuous search) with the exclusion of domains, the steepest descent method with a variable step, a method based on a combination of gradient and random search [9 - 11] have been considered.

Grid search method with the exclusion of domains was used with the objective function (see equation 3) without iterative refinement with the constant penalty parameter r_k = 1,000,000 and parameter s, which is equal to the mean of function $V(\bar{x})$ beyond the boundaries of feasible solutions. Search domain is divided into 4 parts along each of the coordinates.

Grid search method with the exclusion of domains has proved to be a reliable one for searching an optimal solution; it also requires a small number of calculations (384 calculations of the objective function were done in the problem considered) when searching is carried put with a small number of design variables. However, with an increase in the number of design variables a significant increase in the number of calculations of the objective function should be expected which hinders the use of this method in such cases.

Application of the steepest descent method with a variable step and parameter s=0 (objective function without a barrier on the boundary of feasible solutions), has shown that in most cases it does not lead to finding an optimal solution. This is due to a ravine surface shape of the objective function (see equation 3). The algorithm usually 'loops' on the boundary of permissible values domain.

The method, based on a combination of gradient and random search, proved to be a reliable one when solving problems when in equation 3 s=0. To obtain sufficiently accurate results 412 calculations of the objective function were done. This value is higher than the one, which was done for the grid search method with the exclusion of domains. However, with an increase in the number of design variables, in contrast to the grid search method, there is no significant increase in the number of calculations of the objective function. Therefore, it can be recommended for solving problems with the use of the developed optimal design method with a large number of design variables.

6. Conclusions

Methodology of optimal design of geometrically nonlinear shells of revolution of minimum weight with strength, stability and strain constraints is developed. Stress and strain state of the shell is determined by Galerkin method in the mixed finite element definition.

Based on this methodology, algorithms for determining the optimal form and thickness distribution of a shell of minimum weight are obtained. The results of solving two test problems confirmed the possibility and accuracy of finding the optimal solution.

The analysis of the effectiveness of different optimization algorithms to solve the set problem is given. The selected algorithm based on a combination of gradient and random search, allows to effectively searching the optimal solution for a large number of design parameters.

7. Acknowledgement

This work was supported by the grant of the President of the Russian Federation #MK-9203.2016.8

8. References

[1] Fletcher C A J 1984 *Computational Galerkin methods* (Berlin, Heidelberg, New York, Tokyo: Springer-Verlag)
[2] Strang G, Fix G 2008 *An Analysis of the Finite Element Method* (Wellesley-Cambridge Press)
[3] Stupishin L U, Nikitin K E 2014 *Applied Mechanics and Materials* **919-921** 1299-1302
[4] Stupishin L U, Kolesnikov A G 2014 *Applied Mechanics and Materials* **501-504** 766-9
[5] Stupishin L U, Nikitin K E 2014 *Applied Mechanics and Materials* **988** 338-341
[6] Misvhke C R 1968 *An Introduction to Computer-Aided Design* (Englewood Cliffs, N.J, Prentice-Hall)
[7] Andreev V I, Dubrovskiy I A 2013 *Applied Mechanics and Materials* **405-408** 1073-6
[8] Floudas C A , Pardalos P M 2009 *Encyclopedia of Optimization* (New York: Springer Publ.)
[9] Stupishin L U, Kolesnikov A G 2017 *Asian journal of civil engineering (BHRC)* **18** 163-71
[10] Stupishin L U, Kolesnikov A G 2014 *Applied Mechanics and Materials* **580-583** 3062-5
[11] Skokov V A 1977 *Mathematical Methods for Studying Economic Problems* **7** 51-8

IOP Conf. Series: Materials Science and Engineering **311** (2018) 012008 doi:10.1088/1757-899X/311/1/012008

Scheduling of hybrid types of machines with two-machine flowshop as the first type and a single machine as the second type

Ming-Chih Hsiao[1], **Ling-Huey Su**[2]

1, 2 Department of Industrial and Systems Engineering, Chung Yuan Christian University, Taiwan.
E-mail: t6817361@cycu.org.tw

Abstract. This research addresses the problem of scheduling hybrid machine types, in which one type is a two-machine flowshop and another type is a single machine. A job is either processed on the two-machine flowshop or on the single machine. The objective is to determine a production schedule for all jobs so as to minimize the makespan. The problem is NP-hard since the two parallel machines problem was proved to be NP-hard. Simulated annealing algorithms are developed to solve the problem optimally. A mixed integer programming (MIP) is developed and used to evaluate the performance for two SAs. Computational experiments demonstrate the efficiency of the simulated annealing algorithms, the quality of the simulated annealing algorithms will also be reported.

1. Introduction

Most studies on the hybrid machine scheduling consider all n jobs to be processed on a hybrid system that consists of a series of production stages or workshops and at least of which must have several facilities [1]. We consider the different machine environments in which two production machine types are considered and each job instead of being processed on two machine types. It is processed either on the two-machine flowshop or on the single machine. Gupta and Tunc [2] considered a n jobs, two-stage hybrid flowshop scheduling problem with only one machine at stage 1 and m identical machines at stage 2, they developed approximate algorithms to find a minimum makespan. Wang, Liu and Chu [3] investigated a two-stage no-wait hybrid flowshop scheduling problem in which the first stage contains a single machine and the second stage contains several identical parallel machines, in the paper, they proposed a branch-and-bound algorithm to find minimum makespan. Rabiee, Rad, Mazinani and Shafaei [4] proposed a two-stage no-wait hybrid flowshop scheduling problem considering unrelated parallel machines, sequence-dependent setup times, and they used a hybrid algorithm which is combined with simulated annealing, variable neighborhood search and genetic algorithm to search minimum makespan. The hybrid flowshop problems are extended by considering various process conditions. Luo, Huang, Zhang, Dai and Chen [5] considered a two-stage hybrid batching flowshop scheduling with blocking and machine availability constraints. Oğuz, Ercan, Cheng and Fung [6] addressed a heuristic algorithm for a two-stage flowshop scheduling problem with multiprocessor tasks by considering a multi-stage setting for the processor environment to minimize the makespan. Jungwattanakit, Reodecha, Chaovalitwongse and Werner [7] considered a hybrid flowshop with unrelated parallel machines in one of stages. The objective is to minimize a convex sum of makespan and the number of tardy jobs. Rossi, Pandolfi and Lanzetta [8] analysed a multiple-stage hybrid flowshop with sequence-independent uniform set-up time, parallel batching machines with compatible parallel batch families. They used a critical ratio setup rule and a sliding time window technique to develop two heuristics to reduce the tardy jobs and the makespan.

2. Problem description

Using the standard three-filed notation, the problem is denoted as $F2,1||C_{max}$, where $F2$ indicates two-machine flowshop and 1 means a single machine. The jobs can be processed on either the flowshop or the single machine. There are n jobs to be processed either on the two-machine flowshop or on the single machine. The objective is to minimize the makespan.

Assumptions are made in this paper:

- Each job is independent. It is always processed without defect and its machine setup time can be negligible.
- Each job has neither preemption nor priority values for processing in each machine type.
- At any given time, a machine can process only one job. All machines are capable to process all jobs.
- The transportation time of jobs between the machines can be negligible on flowshop machines environment.
- The processing of jobs cannot be interrupted, i.e. once a machine starts processing a job, the job has to be completed without interruptions.
- During scheduling period, there is no maintenance or breakdowns in the machines.

3. Simulated annealing algorithm

The simulated annealing (SA) algorithm, introduced by Kirkpatrick [9], is a local search procedure capable of escaping from local optimum to solve combinatorial optimization problems. To start the procedure, SA draws an initial solution to generate neighborhood solution. If the neighborhood solution is better than the incumbent solution, the former is automatically accepted and replaces the latter; otherwise, the incumbent solution is used. The whole process is repeated until no significant improvement in the neighborhood solution is found or the pre-specified conditions are met. SA uses this repetitive improvement approach, but in particular it enables a search algorithm to escape from a local optimum.

3.1 Initial job sequence

To acquire a good initial solution for the proposed SA, the following steps were used: (1) generate a job sequence randomly; (2) sequence the jobs by Johnson's rule; (3) assign the jobs to either the flowshop or the single machine, which offers the lowest completion time; (4) and let the solution (C_{max}) as the initial solution.

3.2 Decoding method

The solutions of assignment problems are represented by permutation schemes. Two machine types are available to perform the jobs. These jobs are then assigned to the machines and sequence at each machine type in order to minimize the overall makespan. In here, two decoding methods for the proposed SA are recommended. The first decoding method was overall-permutation decoding (OPD). Since the solution is represented by the permutation of the job indices $\{1,2,\cdots,n\}$ and two machine types, hence, there are $(n+2-1)!$ permutations, which stand for the overall solutions. The permutation representation expresses the positions of jobs and the order sequence in the machines. In the permutation schemes, $(n+1)$ was utilized as the partition number to separate the assignment for each machine types. Prior to the partition number $(n+1)$, there are k jobs $(k \in n)$, where k jobs are assigned to the flowshop machines while the remaining jobs are assigned to the single machine. For the second decoding method, a list scheduling with Johnson's rule decoding (LSJRD) was applied. The solution is represented by the permutation of the job indices$\{1,2,\cdots,n\}$. First, all jobs in Johnson's rule were sequenced while the permutation representation displayed the assignment sequence of jobs. The jobs were assigned to the machine types with the lowest completion time.

3.3 Stopping criterion

The stopping criteria are the conditions to complete the algorithm. Theoretically, the SA finishes when the temperature converges to T_F or the solution does not improve after some iterations. In our research, if the temperature is lower than T_F, then the SA will stop.

4. Computational experiments

The experiments are divided into two parts. The first part gives the mixed integer programming optimal solution, which can be used to compare with the proposed SA algorithms as a basis to prove the effectiveness of the SA algorithms. The second part provides the SA algorithms solution. The processing time of flowshop machine is uniformly distributed [1, 25] and the processing time of single machine is between
[max (the processing time of flowshop machines), sum of processing time of flowshop machines]. The numbers of job $n =$
$(6,7,8,9,10,11,12,13,14,15,16,17,18,19,20,30,40,50,60,70,80,90,100,150,200,300)$. The numbers of job generated five different problem instances, which was run thirty times for every instance. Thus, 130 problems were tested in the experiment. For each instance, the optimum makespan and average computation time were obtained and computed from thirty trials.

5. Computational results

This section describes the computational tests to evaluate the performance of the proposed SAs. For small problems, the same test instances were used to run in CPLEX and SAs. The LSJRD and OPD generated 25 optimal solutions (100%) and 24 optimal solutions (96%) for 25 testing problems, respectively (Table 1). The solution performances for two SAs are likely similar with those of the MIP. In addition, the computation time (seconds) needed by the SA is shorter than MIP.

Table 1. Optimal solutions obtained by MIP, OPD and LSJRD.

	MIP		LSJRD		OPD	
Instance #	Optima	Time(sec)	Solution	Time(sec)	Solution	Time(sec)
6.1	28*	2.26	28*	0.011	28*	0.0106
6.2	43*	4.11	43*	0.0081	43*	0.0053
6.3	47*	3.61	47*	0.011	47*	0.0097
6.4	60*	4.77	60*	0.0093	60*	0.0053
6.5	46*	2.06	46*	0.0119	46*	0.0107
7.1	45*	18.164	45*	0.0129	45*	0.0087
7.2	70*	15.256	70*	0.0097	70*	0.0107
7.3	54*	29.166	54*	0.011	55	0.0111
7.4	49*	20.46	49*	0.0117	49*	0.0106
7.5	53*	9.97	53*	0.0113	53*	0.0087
8.1	64*	228.164	64*	0.0232	64*	0.0082
8.2	50*	54.584	50*	0.0223	50*	0.0097
8.3	49*	51.484	49*	0.0224	49*	0.0106
8.4	54*	17.392	54*	0.0221	54*	0.0087
8.5	53*	116.102	53*	0.0231	53*	0.0092
9.1	64*	424.946	64*	0.0171	64*	0.0121
9.2	50*	556.012	50*	0.0218	50*	0.0112
9.3	61*	798.236	61*	0.0213	61*	0.0082
9.4	64*	1792.734	64*	0.0168	64*	0.0106
9.5	55*	625.054	55*	0.0206	55*	0.0092
10.1	88*	7200.24	88*	0.0145	88*	0.0092
10.2	57*	4210.084	57*	0.0167	57*	0.0087
10.3	61*	6376.158	61*	0.0161	61*	0.0073
10.4	62*	4607.54	62*	0.0197	62*	0.0131
10.5	59*	5791.472	59*	0.017	59*	0.0092

To compare two decoding methods for large problems, the decoding methods were run for 130 instances and the best makespan and average computational time were obtained using two decoding methods (OPD and LSJRD). The results were compared for large problems by counting the number of success run for each method. Evidently, the method with larger success run numbers produced better results than the others. Table 2 shows that the comparisons of the two solutions secured by OPD and LSJRD. When the size of the job is small, the OPD and LSJTD have similar performances, however, as the numbers of jobs increase, the performances of OPD decrease. This may be attributed to the obtained number of better solutions from the OPD, which is less than the LSJRD.

Table 2. The solution comparisons for OPD and LSJRD.

LSJRD	6	7	8	9	10	11	12	13	14	15	16	17	18
Better[1]	0	1	0	0	0	0	0	2	1	2	0	4	2
Equal[2]	5	4	5	5	5	5	5	3	4	3	5	1	2
Worse[3]	0	0	0	0	0	0	0	0	0	0	0	0	1

LSJRD	19	20	30	40	50	60	70	80	90	100	150	200	300
Better	4	3	3	2	2	3	3	1	4	2	4	3	3
Equal	1	1	1	2	1	1	0	0	0	1	0	0	0
Worse	0	1	1	1	2	1	2	4	1	2	1	2	2

*1: The result of LSJRD is better than OPD. 2: The result of LSJRD is equal to OPD. 3: The result of LSJRD is worse than OPD.

Furthermore, we applied the average of % improvement from the SAs as compared to the initial solutions. As shown in Table 3, the obtained average % improvement for LSJRD demonstrated better results than the OPD. The computation time in seconds needed by the OPD and LSJRD are shown in Table 4.

Table 3. The average of % improvement from OPD and LSJRD.

	6	7	8	9	10	11	12	13	14
LSJRD	5.39	13.33	12.42	9.69	8.30	9.74	7.96	14.23	10.70
OPD	5.39	12.97	12.42	9.69	8.30	9.74	7.96	13.82	10.16

	15	16	17	18	19	20	30	40	50
LSJRD	13.55	8.80	10.46	11.77	10.46	9.37	8.97	8.15	7.44
OPD	12.82	8.80	9.60	11.11	9.54	9.06	8.20	7.79	7.46

	60	70	80	90	100	150	200	300
LSJRD	7.68	6.79	5.28	6.29	6.09	6.24	4.59	4.72
OPD	7.16	6.35	5.65	6.06	5.87	5.77	4.38	4.57

Table 4. The average computation time for OPD and LSJRD.

	6	7	8	9	10	11	12	13	14
LSJRD	0.0103	0.0113	0.012	0.0142	0.0168	0.0195	0.0226	0.0242	0.0294
OPD	0.0083	0.01	0.0093	0.0103	0.0095	0.0116	0.0115	0.0125	0.0142

	15	16	17	18	19	20	30	40	50
LSJRD	0.0319	0.0353	0.04	0.043	0.0472	0.0508	0.1117	0.2003	0.3201
OPD	0.0146	0.015	0.0164	0.0165	0.0163	0.0171	0.027	0.0397	0.049

	60	70	80	90	100	150	200	300
LSJRD	0.4785	0.6877	0.9518	1.2754	1.7412	5.383	11.876	36.617
OPD	0.0641	0.0777	0.0931	0.1109	0.132	0.2497	0.4224	0.825

6. Conclusions

In this paper, we addressed the hybrid machine types scheduling problem, in which one type is a two-machine flowshop and the other type is the single machine. The problem was inspired by the IC final testing operation in IC backend factory. Two SAs were proposed to effectively respond on the large problems by applying the optimal solution for MIP model. A remarkable initial solution was achieved by adopting the efficient initial solution for the proposed SAs. Besides, the two decoding methods were applied to generate the total completion time. From the computational experiments, the optimal solution of small size problems for LSJRD and OPD were 100% and 96%, respectively. The average computation times for the two SAs were extremely shorter to that of the MIP. For large size problems, LSJRS demonstrated better than OPD, which was based on the obtained number of better solutions. Furthermore, the former exhibited better average of % improvement than the latter. It was found that the average computation times of LSJRD started to increase (i.e over 1 second) after 90 jobs. Based on the actual factory environment, 36.6 seconds are needed for 300 jobs, which offer an equitable and acceptable time. This paper shows a new solution on scheduling hybrid machine types, wherein one type is a two-machine flowshop and the other is the single machine, for NP-hard scheduling issue. The proposed SAs proved its capacity in obtaining the optimal solutions efficiently. In the future, we will extend the application of the proposed SAs to hybrid types of machines types with buffer constrains. Also, our research will continue designing meta-heuristics for the problem by taking advantage of good initial solutions generated by the presented heuristics.

7. References

[1] Gupta J N 1988 Two-stage, hybrid flowshop scheduling problem *Journal of the Operational Research Society* 359-64

[2] Gupta J N and Tunc E A 1991 Schedules for a two-stage hybrid flowshop with parallel machines at the second stage *International Journal of Production Research* **29** 1489-502

[3] Wang S, Liu M and Chu C 2015 A branch-and-bound algorithm for two-stage no-wait hybrid flow-shop scheduling *International Journal of Production Research* **53** 1143-67

[4] Rabiee M, Rad R S, Mazinani M and Shafaei R 2014 An intelligent hybrid meta-heuristic for solving a case of no-wait two-stage flexible flowshop scheduling problem with unrelated parallel machines *International Journal of Advanced Manufacturing Technology* **71** 1229-45

[5] Luo H, Huang G Q, Zhang Y F, Dai Q Y and Chen X 2009 Two-stage hybrid batching flowshop scheduling with blocking and machine availability constraints using genetic algorithm *Robotics and Computer-Integrated Manufacturing* **25** 962-71

[6] Oğuz C, Ercan M F, Cheng T E and Fung Y-F 2003 Heuristic algorithms for multiprocessor task scheduling in a two-stage hybrid flow-shop *European Journal of Operational Research* **149** 390-403

[7] Jungwattanakit J, Reodecha M, Chaovalitwongse P and Werner F 2009 A comparison of scheduling algorithms for flexible flowshop problems with unrelated parallel machines, setup times, and dual criteria *Computers & Operations Research* **36** 358-78

[8] Rossi A, Pandolfi A and Lanzetta M 2014 Dynamic set-up rules for hybrid flowshop scheduling with parallel batching machines *International Journal of Production Research* **52** 3842-57

[9] Kirkpatrick S, Gelatt C D and Vecchi M P 1983 Optimization by simulated annealing *science* **220** 671-80

IOP Conf. Series: Materials Science and Engineering **311** (2018) 012009 doi:10.1088/1757-899X/311/1/012009

Quality control process improvement of flexible printed circuit board by FMEA

Siwaporn Krasaephol[1] and Parames Chutima[1,2*]

1Department of Industrial Engineering, Faculty of Engineering, Chulalongkorn University, Bangkok, Thailand 10330
2 Regional Centre for Manufacturing Systems Engineering, Faculty of Engineering, Chulalongkorn University, Bangkok, Thailand 10330
E-mail: 1 siwaporn.kr@student.chula.ac.th, 2 cparames@chula.ac.th
* Corresponding author

Abstract. This research focuses on the quality control process improvement of Flexible Printed Circuit Board (FPCB), centred around model 7-Flex, by using Failure Mode and Effect Analysis (FMEA) method to decrease proportion of defective finished goods that are found at the final inspection process. Due to a number of defective units that were found at the final inspection process, high scraps may be escaped to customers. The problem comes from poor quality control process which is not efficient enough to filter defective products from in-process because there is no In-Process Quality Control (IPQC) or sampling inspection in the process. Therefore, the quality control process has to be improved by setting inspection gates and IPCQs at critical processes in order to filter the defective products. The critical processes are analysed by the FMEA method. IPQC is used for detecting defective products and reducing chances of defective finished goods escaped to the customers. Reducing proportion of defective finished goods also decreases scrap cost because finished goods incur higher scrap cost than work in-process. Moreover, defective products that are found during process can reflect the abnormal processes; therefore, engineers and operators should timely solve the problems. Improved quality control was implemented for 7-Flex production lines from July 2017 to September 2017. The result shows decreasing of the average proportion of defective finished goods and the average of Customer Manufacturers Lot Reject Rate (%LRR of CMs) equal to 4.5% and 4.1% respectively. Furthermore, cost saving of this quality control process equals to 100K Baht.

1. Introduction

FPCB is components in the smart phone, it is produced by electronics manufacturer. A study factory is one of the FPCB manufacturer, its production begins from receiving FPCBs in sheet form from suppliers then bringing them into punching and assembly processes. At the end of the assembly process, FPCBs will be cut from sheet form to single pieces for 100% checking functions at the In-Circuit Test (ICT) and Functional In-circuit Test (FCT), respectively. Then FPCBs will be 100% inspected appearances at the final inspection process and sampling inspection at the Out-going Quality Assurance (OQA) inspection process. Proportion of defective finished goods were found at the final inspection process have to be scraped because could not be reworked. It made high scrap cost. Often, finished goods do not achieve production capacity's target per day. Production data for FPCBs, model 7-Flex from June 2016 to May 2017 shows the average proportion of defective finished goods, OQA

IOP Conf. Series: Materials Science and Engineering **311** (2018) 012009 doi:10.1088/1757-899X/311/1/012009

Lot Reject Rate (%LRR of OQA), and Customer Manufacturers Lot Reject Rate (%LRR of CMs) equal to 5.7%, 8.4% and 5.9%, respectively and the scrape cost is around 1,000K Baht.

2. Problems

Focusing on FPCB's production, model 7-Flex makes researcher to realize that the root cause of high defective proportion at the final inspection process because of the poor quality control process in current. It is not efficient enough to filter defective products from in-process because there is no IPQC or sampling inspection for appearance inspection during the process. Currently, the appearance inspection process have only one position but two inspection gates. There are the final inspection to 100% inspect and OQA inspection to sampling by following AQL 0.4, level II. Both of them are located at the end of assembly process before auto packing process. These are the minimum for appearance inspection of customer requirement. The current manufacturing and inspection processes for FPCB, model 7-Flex are shown in 'Fig. 1'. Although inspectors at the final inspection processes rejected and removed defective finished goods but they were also found at the OQA inspection process and escaped to customers. Top three of defective finished goods were found at the inspection process from June 2016 to May 2017 are namely wrinkle, scratch and dent on Anisotropic Conductive Film (ACF) areas, their defective proportion equal to 2.8%, 1.6% and 0.4%, respectively. Defective photos are shown in Table 1.

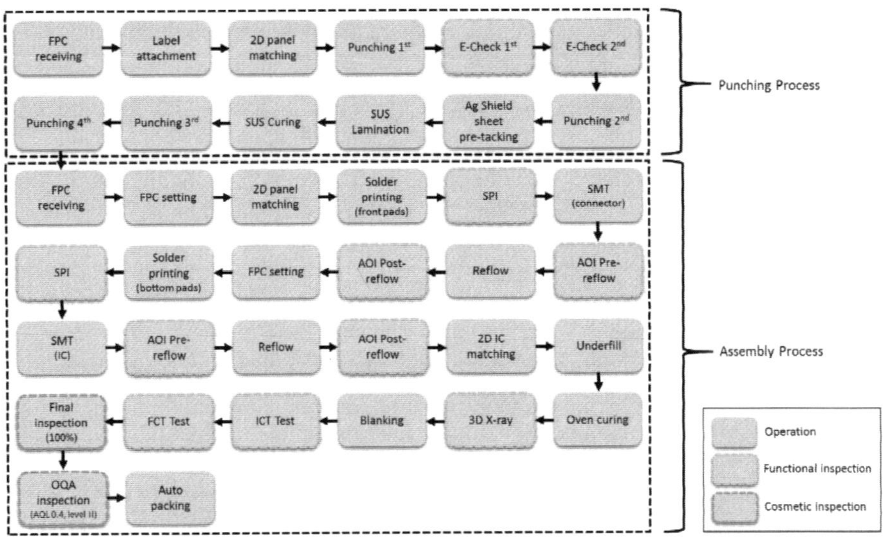

Figure 1. Manufacturing and inspection processes.

Table 1. Top three defective.

Top three defective	1st	2nd	3rd
Photos	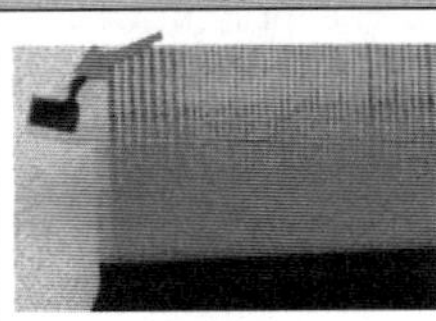		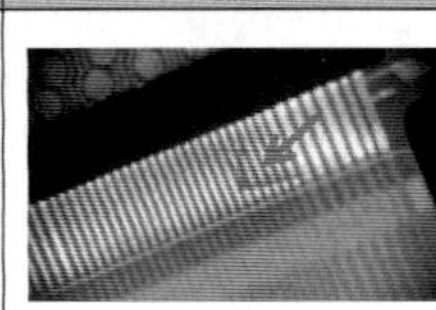
Defective type	Wrinkle on ACF	Scratch on ACF	Dent on ACF
Defective proportion	2.80%	1.60%	0.40%

3. Analysis procedures

3.1. Analyze defective finished goods

Top three defects from June 2016 to June 2017 are namely wrinkle, scratch and dent on Anisotropic Conductive Film (ACF) areas, respectively. All of them is defects caused by the FPCBs movement. Every process that has FPCB movement activities also has chances to generate defects on products. Actually, defective proportion from June 2016 to May 2017 is around 5.7%. Pareto chart of defective proportion is shown in 'Fig. 2'. If researcher applies Pareto's rule, the proportion of wrinkle on ACF is more than 20% of all defective units so it is priority that have to be detected and removed from processes before loading FPCBs into assembly process because they do not have components. Scrap cost of PFCB without component is lower than FPCBs with components. So if the defective products are removed from processes before the assembly process, they will make cost saving. Moreover, the Pareto chart shows some defects were escaped form supplier such as abnormal plating, incomplete gold plating and air bubble. Due to these defects cannot be completely generated by 7-Flex's production.

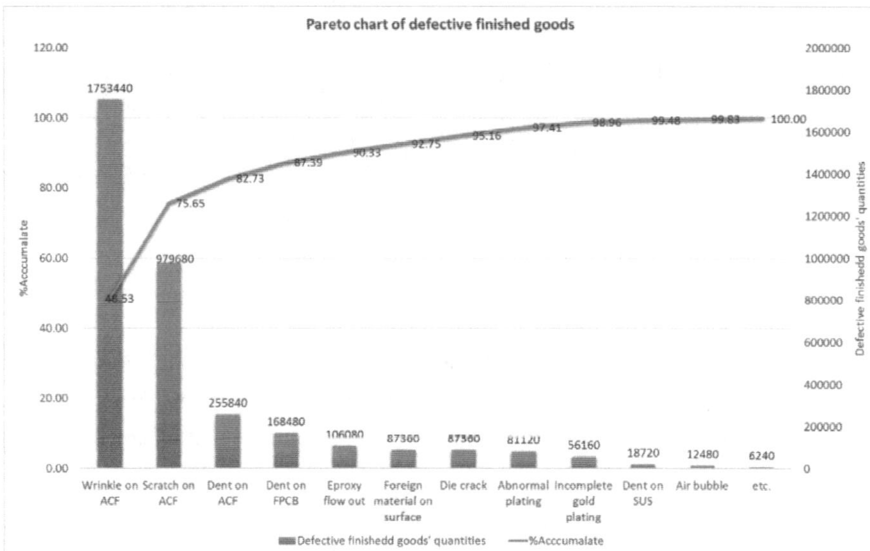

Figure 2. Pareto chart of defective finished goods.

3.2. Analyze processes

Process analysis is step for finding processes that can generate defects on products both functions and appearances. Researcher and production team considered together by using (FMEA method for analyzing to determine risks of failure modes of each process. Scoring for severity, occurrence, and detection of failure modes are shown in Table 2. The FMEA's result of current 7-Flex's punching and assembly processes are shown in Table 3 and Table 4, respectively. The first 20% of Risk Priority Number (RPN) following Pareto's rule is namely punching 1st, punching 2nd, punching 3rd, punching 4th, blanking, and auto packing process, respectively.

Table 2. Scoring of FMEA.

Severity of effect on products (customer effect)	Rank	Occurrence of failure modes	Rank	Ddetection by design control	Rank
All finished goods have to be scraped and production line have to be stopped.	10	Defective rate is more than 10%	10	Failure modes can not be detected.	10
All finished goods have to be scraped but production line have not to be stopped.	9	Defective rate is between 5% to 10%	9	Failure modes are difficult to detect.	9
All finished goods are waived before shipment.	8	Defective rate is between 1.50% to 4.99%	8	Failure modes can be detected by operators' self-check.	8
Finished goods >50% have to be scraped but remaining units can br reworked.	7	Defective rate is between 0.30% to 1.49%	7	Failure modes can be detected by operators' self-check during process.	7
Finished goods <=50% have to be scraped but remaining units can br reworked.	6	Defective rate is between 0.10% to 2.99%	6	Failure modes can be detected by Go-No-Go fixtures.	6
All finished goods can br reworked.	5	Defective rate is between 0.05% to 0.09%	5	Failure modes can be detected by auto machine.	5
All finished goods have defects but can shipment in terms of Use As If (UAI).	4	Defective rate is between 0.01% to 0.04%	4	Failure modes can be detected by Go-No-Go fixtures during process.	4
All finished goods have defects but can shipment because they are within loose specification	3	Defective rate is between 0.0001% to 0.0099%	3	Failure modes can be detected by auto machine during process.	3
All finished goods have defects but can shipment because they are within normal specification.	2	Defective rate is less than 0.0001%	2	Failure modes can be detected by auto machine and process is locked by interlock system.	2
No defective finished goods.	1	Failure modes are protected.	1	Failure modes can be detected by auto machine during process and process is locked by interlock system.	1

Table 3. Result of FMEA of punching process.

No.	Sub-process	Potential failure modes	Severity	Occurrence	Detection	RPN
1	FPC receiving	Handling defects (Dent, Scratch, Wrinkle)	6	8	8	384
		Escape defects from supplier	6	8	8	384
2	Label attachment	Handling defects (Dent, Scratch, Wrinkle)	6	8	8	384
		Contamination	5	7	8	280
3	2D panel matching	Bad mark programming error cause missing components	6	6	5	180
4	Punching 1	Handling defects (Dent, Scratch, Wrinkle)	6	8	8	384
		Dent from blanking die	9	7	8	504
		Blanking misalingment	7	5	7	245
5	E-Check 1	Handling defects (Dent, Scratch, Wrinkle)	6	8	8	384
		Dent from tooling press	9	4	8	288
		Eacape fail E-check	9	4	9	324
6	E-Check 2	Handling defects (Dent, Scratch, Wrinkle)	6	8	8	384
		Dent from tooling press	9	4	8	288
		Eacape fail E-check	9	4	9	324
7	Punching 2	Handling defects (Dent, Scratch, Wrinkle)	6	8	8	384
		Dent from blanking die	9	7	8	504
		Blanking misalingment	7	5	7	245
8	Ag shield sheet applied	Handling defects (Dent, Scratch, Wrinkle)	6	8	8	384
		Ag shield misalingment	7	4	7	196
		Contamination	5	6	8	240
		Fiber	5	3	8	120
		Foreign material inside	9	3	8	216
		Air bubble, Delamination	5	4	7	140
9	SUS Lamination	Handling defects (Dent, Scratch, Wrinkle)	6	8	8	384
		SUS misalignment	9	3	7	189
		SUS dent	9	4	8	288
		SUS reverse	9	3	8	216
10	SUS curing	Handling defects (Dent, Scratch, Wrinkle)	6	8	8	384
		SUS peel	9	2	7	126
11	Punching 3	Handling defects (Dent, Scratch, Wrinkle)	6	8	8	384
		Dent from blanking die	9	7	8	504
		Blanking misalingment	7	5	7	245
12	Punching 4	Handling defects (Dent, Scratch, Wrinkle)	6	8	8	384
		Dent from blanking die	9	7	8	504
		Blanking misalingment	7	5	7	245

Table 4. Result of FMEA of assembly process.

No.	Sub-process	Potential failure modes	Severity	Occurrence	Detection	RPN
13	FPC receiving	Handling defects (Dent, Scratch, Wrinkle)	6	8	8	384
		Air bubble, Delamination	5	4	7	140
		Pad peel off, Ink peel off	9	2	7	126
14	FPC setting	Handling defects (Dent, Scratch, Wrinkle)	6	8	8	384
15	2D panel matching	Bad mark programming error cause missing components	9	5	5	225
16	Solder printing	Solder defects (Height, Volume, Shape)	5	3	5	75
		Printing misalignment	9	3	5	135
17	Solder Printing Inspection (SPI)	SPI Programming error cause over reject	1	5	8	40
18	SMT on connector side	Component misalignment	9	3	8	216
		Wrong component	9	3	8	216
19	AOI Pre-reflow	AOI Programming error cause over reject	1	2	8	16
20	Reflow	Solder defects (Cosmetic cause from heat)	1	2	8	16
		Over heat	9	2	8	144
21	AOI Post-reflow	AOI Programming error cause over reject	1	2	8	16
22	FPC setting	Handling defects (Dent, Scratch, Wrinkle)	6	8	8	384
23	Solder printing	Solder defects (Height, Volume, Shape)	5	3	5	75
		Printing misalignment	9	3	5	135
24	Solder Printing Inspection (SPI)	SPI Programming error cause over reject	1	5	8	40
25	SMT on IC side	Component misalignment	9	3	8	216
		Wrong component	9	3	8	216
		Porality reverse	9	2	8	144
26	AOI Pre-reflow	AOI Programming error cause over reject	1	2	8	16
27	Reflow	Solder defects (Cosmetic cause from heat)	1	2	8	16
		Over heat	9	2	8	144
28	AOI Post-reflow	AOI Programming error cause over reject	1	2	8	16
29	2D IC matching	Bad mark programming error cause missing data	1	5	5	25
30	Underfill dispensing	Epoxy defects over flow	6	6	9	324
		Contamination on IC surface	6	6	7	252
31	Oven curing	Handling defects (Dent, Scratch, Wrinkle)	6	8	8	384
		Un-curing epoxy glue	5	2	7	70
		Epoxy crack	4	2	8	64
32	3D X-Ray	Handling defects (Dent, Scratch, Wrinkle)	6	8	8	384
33	Blanking	Handling defects (Dent, Scratch, Wrinkle)	6	8	8	384
		Dent from blanking die	9	7	8	504
		IC crack	9	6	8	432
		Blanking misalingment	7	5	7	245
34	In-Circuit Test (ICT)	Handling defects (Dent, Scratch, Wrinkle)	6	8	8	384
		Dent from fixture or pin in machine	9	4	8	288
35	Functional Circuit Test (FCT)	Handling defects (Dent, Scratch, Wrinkle)	6	8	8	384
		Dent from fixture or pin in machine	9	4	8	288
36	Final inspection	Eacape defects	1	3	9	27
37	OQA inspection	Eacape defects	1	2	9	18
38	Auto packing	Handling defects (Dent, Scratch, Wrinkle)	1	2	10	20

3.3. Improve current quality control processes

The defective finished goods analysis makes researcher to know that defective proportion of raw materials were escaped to 7-Flex's production and found at the final inspection process up to 0.3% from 5.7%. But in fact, production accepts defective proportion from raw material only 0.25%. So researcher designs inspection gate to sampling raw material before bringing production, this gate is called In-coming Quality Assurance (IQA) inspection process. MIL-STD-105E is applied for this gate because inspection's results are attribute data. Single sampling plan is used because it is the easiest for inspectors to use. This gate is designed to follow AQL 0.25, general inspection level II because this inspection is not destruct products and products' price is not too expensive. Furthermore, the FMEA's result shows the maximum and minimum score of RPN equal to 504 and 16, respectively. Inspection gates should be added after the critical processes for fastest detecting and filtering defective products during process. Practically, from punching 1st to punching 4th process is continuous process and FPCBs are operated in sheet form. So researcher adds the inspection gate next to punching 4th process to 100% inspect and individual bad mark the defective products, this gate is called punching inspection process. Surface Mount Technology (SMT) process at assembly process will not mount component on bad marked products. Another inspection gate is added next to punching inspection to assure product's quality by sampling because the punching inspection process and its inspectors are new. Single sampling plan is still used for this gate by following AQL 0.4, level II because target of defective proportion in-process is not over 0.5%. This gate is called punching Out-going Quality Assurance (OQA) inspection process. After finished the punching process, FPCBs will be brought into assembly process. FPCBs are still operated in sheet form by fixtures and removed from the fixture after finished 3D X-Ray process. From FPC receiving to 3D X-Ray process at the assembly process is continuous processes and uses the same fixture. The FMEA's result shows RPN of epoxy overflow defect of under-fill dispensing process equals to 324. It is too high because if epoxy is stained on IC's surface and cured at oven curing process (next to the under-fill dispensing process), FPCB have to be scraped. But if epoxy on IC's surface is cleaned before curing, FPCB can accept. The defective proportion of epoxy overflow equals to 0.2% from 5.7%. Although it is not too much but cannot rework after curing so operators have to 100% inspect for the defect and clean if they find the defect. For this reason, another inspection gate is added next to the under-fill dispensing process to specially check epoxy on IC's surface. Next critical process is blanking process because it have press, cutting, and many movements. After finished the blanking process, FPCB in sheet form becomes single piece and is tested functions at ICF and FCT processes. After that it is inspected appearance at the final inspection process. Even if the blanking process is critical to generate defects, they are detected at the final inspection process. Next to OQA inspection process is auto packing process and not critical. The new inspection processes for 7-Flex's production is shown in 'Fig. 3'. Totally, researcher add more four inspection gates. This is the improved quality control process.

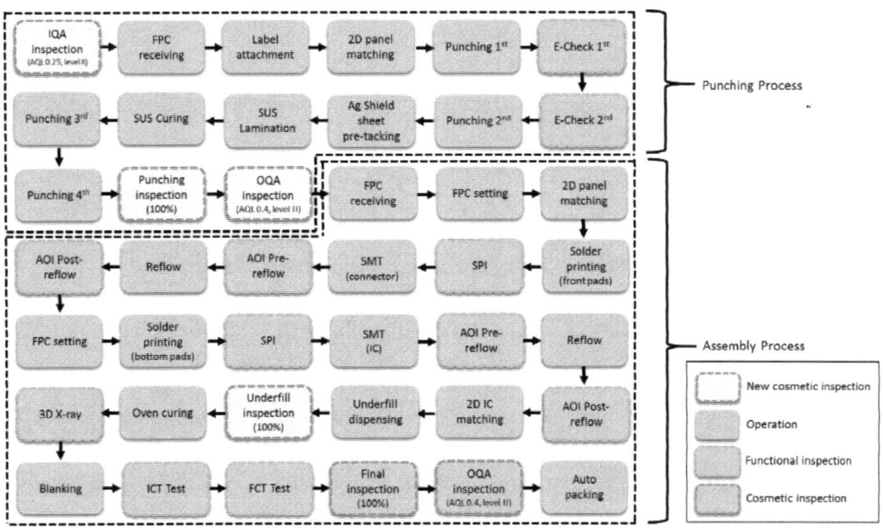

Figure 3. New inspection processes.

3.4. Implement improved quality control processes

The improved quality control process have implemented for FPCB production, focus on model 7-Flex since July 2017 until. Moreover, the improved quality control process is used for production in terms of quality control plan (QCP) document to check processes before starting operations by operators. It identifies process sequences and their parameters that have to be checked include checking method. The parameters are separated by two characteristics, there are process and product. The checking method consists of equipment, sample size, and for checking and specification for referring. Furthermore, the QCP also identifies reaction plan to inform problems if operators find abnormal parameters after checked.

4. Results

After improved the quality control process and have implemented for FPCB production, model 7-Flex from July 2017 until now. Production data of inspection processes from June 2016 to September 2017 is shown in Table 5. It consists of the %LRR of IQA, the defective proportion at punching inspection process, the %LRR of OQA at punching inspection process, the rework rate at under-fill process, the defective proportion at final inspection process, the %LRR of OQA at final inspection process, and the %LRR of CMs. The inspection data of the old inspection processes have trend to decrease after implemented the improved quality control process while the inspection data of the new inspection processes can detect defective products. The average of defective proportion at final inspection process, the average of %LRR of OQA at final inspection process, and the average of %LRR of CMs in last three month equal to 1.1%, 1.0%, and 1.7%, respectively, they decrease from the average data of June 2016 to May 2016 equal to 5, 8, and 3.4 times, respectively. Trends of data of the six inspection processes and data of the customers' inspection process from September 2016 to 2017 are shown in "Fig. 4'. And important, cost saving after the study factory implemented the improved quality control process from July 2017 equals to 100K Baht because the scraps in finished goods level or FPCBs with components at final process are decreased and the scraps cost of FPCBs in sheet form at punching process are less than the scraps cost of FPCBs in finished goods level at assembly process. The scrap cost of finished goods with components is around 300 Baht per unit but the scrap cost of sheet form without components is around 10 Baht per unit.

Table 5. Inspection data of 7-Flex production.

Year	Month	%LRR of IQA (%)	Defective proportion at punching inspection (%)	%LRR of OQA at punching inspection (%)	Rework rate at underfill inspection (%)	Defective proportion at final inspection (%)	%LRR of OQA at final inspection (%)	%LRR of CMs (%)
2016	June	N/A	N/A	N/A	N/A	8.57	30.08	N/A
	July	N/A	N/A	N/A	N/A	7.64	15.33	N/A
	August	N/A	N/A	N/A	N/A	6.35	6.58	N/A
	September	N/A	N/A	N/A	N/A	5.79	5.76	8.45
	October	N/A	N/A	N/A	N/A	4.51	6.39	6.04
	November	N/A	N/A	N/A	N/A	4.32	6.43	7.25
	December	N/A	N/A	N/A	N/A	4.01	5.64	6.82
2017	January	N/A	N/A	N/A	N/A	5.78	5.78	6.49
	February	N/A	N/A	N/A	N/A	5.87	5.32	5.81
	March	N/A	N/A	N/A	N/A	4.77	4.9	4.81
	April	N/A	N/A	N/A	N/A	5.79	4.42	4.5
	May	N/A	N/A	N/A	N/A	4.46	4.08	4.19
	June	N/A	N/A	N/A	N/A	5.12	4.15	4.08
	July	5.19	4.74	3.94	0.52	1.13	1.22	1.81
	August	4.59	4.92	3.00	0.49	0.92	0.91	1.56
	September	5.32	5.01	2.92	0.60	1.12	1.01	1.59

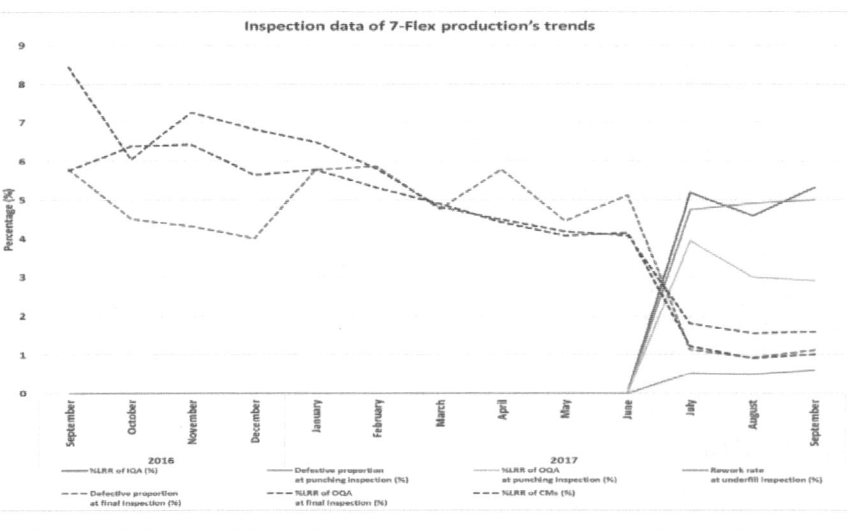

Figure 4. Trends of data of the inspection processes of 7-Flex production.

IOP Conf. Series: Materials Science and Engineering **311** (2018) 012009 doi:10.1088/1757-899X/311/1/012009

5. Conclusions

FMEA method is used for analysing processes to estimate their failure modes' risk then add appearance inspection gates next to the critical manufacturing processes both 100% inspection and sampling. In-process inspection gates are added for checking and reworking product immediately. Totally, four new inspection gates are added, the improved quality control process have total six appearance inspection gates. Inspection's results after implement the new improved quality control from July 2017 are decreased as expected. The defective proportion, %LRR, and scraps at the final inspection process are decrease because some defective products were detected and removed since the punching process and were reworked next to under-fill process. Moreover, decreasing of scraps in finished goods level make cost saving almost 100K Baht

6. References

[1] J. Webber, (1990). "FMEA: Quality Assurance Methodology", Industrial Management & Data Systems, Vol. 90 Issue: 7, pp.21-23.

[2] P. Brennecke, I. Beckmerhagen, (1994). "Quality Assurance and Quality Control for the Planned Konrad Repository", International Journal of Quality & Reliability Management, Vol. 11 Issue: 5, pp.30-37.

[3] K.Y. Kwok, V.M. Rao Tummala, (1998). "A quality control and improvement system based on the total control methodology (TCM)", International Journal of Quality & Reliability Management, Vol. 15 Issue: 1, pp.13-48.

[4] Belmiro P.M. Duarte, Pedro M. Saraiva, (2008). "An optimization - based approach for designing attribute acceptance sampling plans", International Journal of Quality & Reliability Management, Vol. 25 Issue: 8, pp.824-841.

[5] M. Gustavsson, C. Wänström, (2009). "Assessing information quality in manufacturing planning and control processes", International Journal of Quality & Reliability Management, Vol. 26 Issue: 4, pp.325-340.

[6] C. Paciarotti, G. Mazzuto, D. D'Ettorre, (2014). "A revised FMEA application to the quality control management", International Journal of Quality & Reliability Management, Vol. 31 Issue: 7, pp.788-810.

[7] L. Appaia, S. Kandaswamy, (2014). "Selection of single sampling plans by attributes under the conditions of zero-inflated Poisson distribution", International Journal of Quality & Reliability Management, Vol. 31 Issue: 9, pp.1002-1011.

[8] C. Kahraman, E. Turanoglu Bekar, O. Senvar. A Fuzzy Design of Single and Double Acceptance Sampling Plans 179-211.

IOP Conf. Series: Materials Science and Engineering **311** (2018) 012010 doi:10.1088/1757-899X/311/1/012010

Transportation of part supply improvement in agricultural machinery assembly plant

Anusit Saysaman[1] and Parames Chutima[1,2*]
1 Department of Industrial Engineering, Chulalongkorn University, Bangkok, Thailand 10330
2 Regional Centre for Manufacturing Systems Engineering, Faculty of Engineering Chulalongkorn University, Bangkok, Thailand 10330
E-mail: 1Anusit.S@student.chula.ac.th, 2cparames@chula.ac.th
*corresponding author

Abstract. This research focused on the problem caused by the transportation of part supply in agricultural machinery assembly plant in Thailand, which is one of the processes that are critical to the whole production process. If poorly managed, it will affect transportation of part supply, the emergence of sink cost, quality problems, and the ability to respond to the needs of the customers in time. Since the competition in the agricultural machinery market is more intense, the efficiency of part transportation process has to be improved. In this study, the process of transporting parts of the plant was studied and it was found that the efficiency of the process of transporting parts from the sub assembly line to its main assembly line was 83%. The approach to the performance improvement is done by using the Lean tool to limit wastes based on the ECRS principle and applying pull production system by changing the transportation method to operate as milkrun for transportation of parts to synchronize with the part demands of the main assembly line. After the transportation of parts from sub-assembly line to the main assembly line was improved, the efficiency raised to 98% and transportation process cost was saved to 540,000 Baht per year.

1. Introduction

Agricultural machinery assembly plant under this case study has produced and distributed products both domestically and internationally. It is a leading agricultural machinery plant in the industry in Thailand with a market share of more than 80 percent of the entire market. From the intense competition in the agricultural machinery market, the production process efficiency must be improved in order to meet the needs of customers on time. From the study, it was found that the problem of the factory used for the case study was the delivery of parts failed to meet the needs of the main assembly line, which affect the production process efficiency. The delivery process comprised of 6 part types packed into plastic boxes and transported by 3 electric carts as shown in Figure 1: Gear 1, Gear2, Brake Ass'y, Hydraulic Cylinder Ass'y, Lever and RR Axle. The data collected from the problems of delivering parts not meeting the demand on time in April and May 2017 as shown in Table1, it was found that the cause of parts not meeting the demand on time was by delayed delivery of the most as shown in Figure 2 resulting in the mean efficiency of the transportation of parts from sub-assembly lines to the main assembly was 83 %, which was efficient based on the transportation efficiency calculation from the equation.

$$\text{Transportation Efficiency} = \frac{\text{Number of trips to deliver on time}}{\text{Total number of all delivery trips}} \qquad (1)$$

IOP Conf. Series: Materials Science and Engineering **311** (2018) 012010 doi:10.1088/1757-899X/311/1/012010

Figure 1. Electric cart for part supply.

Table 1. Late parts deliver data.

		APRIL		MAY	
		Time	Percentage	Time	Percentage
Part Supply Cycle	Time	1045		1571	
Detail					
Picking Delay	Time	24		37	
Suply Delay	Time	186		254	
Other	Time	3		2	
Part Supply Efficiency	Percentage	82		84	

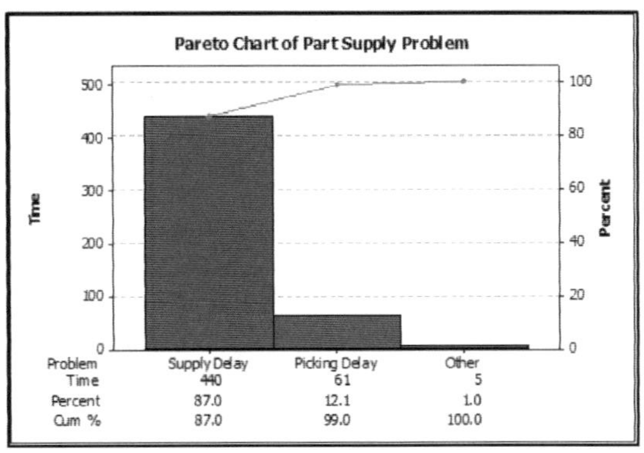

Figure 2. Pareto chart of part supply problem.

2. Problem analysis

From the analysis on the parts transportation in the current Flow Process Chart and a summary of its activities in the transport process as shown in Table 2, it was found that the performance of 3 electric carts drivers cause the wastes in the process as follows.

Table 2. Part transportation process activities.

No.	Flow Process Chart	Activity				
		○	⇨	D	□	▽
1	Brake Ass'y	3	4	-	-	-
2	Gear1, Gear2, Hydraulic Cylinder Ass'y	9	7	-	-	-
3	Lever, RR Axle	11	9	-	-	-
Total		**23**	**20**	**-**	**-**	**-**

2.1 Delivery routes

There are different transportation routes for the 3 electric carts to deliver small parts from the sub-assembly line to the main assembly line. The transportation routes are also not straight, which requires more time to transport. As seen in Table 3 and Figure 3, the transportation distance of the 6 part types was a total of 622 meters. The 1st transport vehicle took 560 seconds to transport Brake Ass'y. The 2nd transport vehicle took 645 seconds to transport Gear1, Gear2, and Hydraulic Ass'y at. The 3rd vehicle took 629 seconds to transport the lever and RR Axle.

Table 3: The transportation distance and time data.

Route	Part for Supply	Distance (meter)	Cycle Time (second)
1	Brake Ass'y	175	560
2	Gear1, Gear2, Hydraulic Cylinder Ass'y	224	643
3	Lever, RR Axle	223	629
Total		622	1832

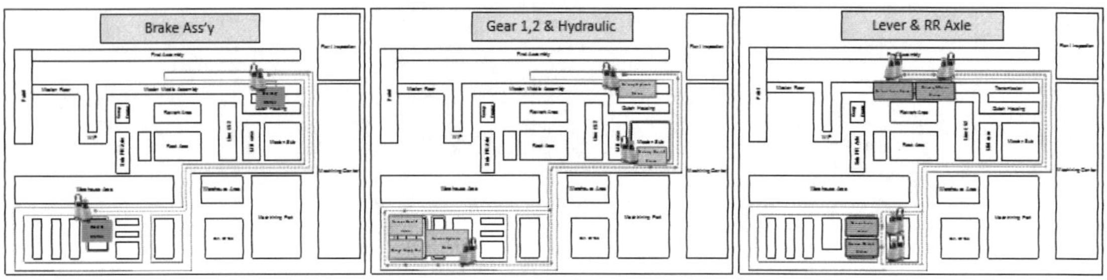

Figure 3. Part supply route.

2.2 Procedures

From the analysis on Flow Process Chart, the procedures required staffs to unload parts and empty containers from vehicles every time were found. The parts and empty container unloading points are in different positions. This makes employees lose more time to work. From the study on the time required to unload parts and empty containers by all 3 drivers as described in Table 4, the process took a total of 442 seconds per 1 transportation from the sub-assembly line to its main assembly line.

Table 4: Time spent to unload parts and empty containers.

Cart	Part	Time (sec)
1	Brake Ass'y	90
2	Gear 1	131
	Gear 2	
	Hydraulic Cylinder Ass'y	
3	Lever	221
	RR Axle	

3. Process improvement

Wastes removal tool based on ECRS and pull system were used to improve the efficiency of the transportation process of agricultural machinery parts for the part transport process to synchronize with the demand of the main assembly line. There were totaling four approaches to the improvements as follows:

3.1. Improving the transport routes

The objective was to reduce the distance between the sub-assembly line to the main assembly line by improving the layout of the sub-assembly line location and defining the receiving point to be facing the road used for the transportation of all the components as shown in Figure 4. So the Milkrun car will pick up the parts in a single route and straight line. This results in the transport routes after the improvement be at a distance of 250 meters, as shown in Figure 5, down from the original 412 meters.

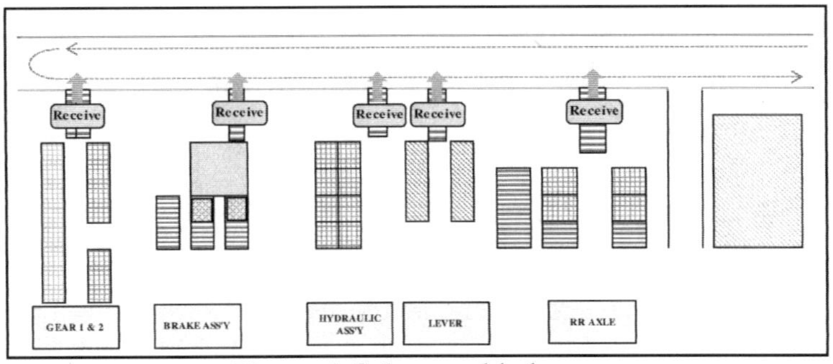

Figure 4. Sub assembly layout.

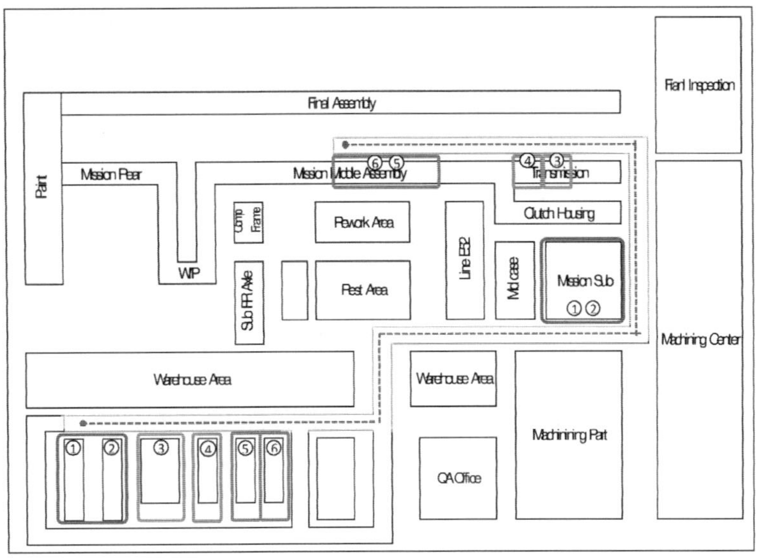

Figure 5. Part supply route for milkrun.

3.2 Improving the parts loading/unloading system

The objective was to reduce the loading/unloading process of return boxes and part boxes with the staffs having to park the vehicle each and every time to load empty containers and drive to the part loading point to load the parts into vehicles (Receive) or the delivery point, which is a waste that occurs in the process. Therefore, the concept of preparation and loading empty containers and part container in the same position was used to solve such problems and it is also automatically loaded as shown in Figure 6. This results in the drivers not having to get out of the vehicle to perform the loading or hooking up with another vehicle. It could reduce the time the process of loading return box and part box from 442 seconds to 150 seconds per transportation.

Figure 6. Auto load system.

IOP Conf. Series: Materials Science and Engineering **311** (2018) 012010 doi:10.1088/1757-899X/311/1/012010

3.3 Improving the transport system to Milkrun

Milkrun vehicle concept is from the need to deliver the required number of parts to synchronize with the main assembly line demand to solve the problem of on time delivery or not exceeding the required number. This affects performance, cost of production in this agricultural machinery factory. The design of the Milkrun principle for the transportation of parts from the sub-assembly line to the main assembly line is to deliver parts that need to be delivered to the destinations that are on the same route to be delivered by the same vehicle. However, with the limitation that not all parts can be placed on the same Milkrun vehicle because of the length of the Milkrun, which affects the efficiency of the control of the vehicle turning into curves. Therefore the vehicles are split into 2 units: the 1st vehicle would transport Gear 1, Gear 2 and Brake Ass'y parts as shown in Figure 7 cars. The 2nd one would transport Lever and RR Axle parts as shown in Figure 8.

Figure 7. Milkrun No.1.

Figure 8. Milkrun No.2.

3.4. Visual control system

This was designed to synchronize a form of parts transport between the vehicles and the main assembly line so the idea of changing transportation system parts from push system a production system to push to pull system to achieve a transport system that automatically meets the needs of customers. If the customer does not pull the parts to assemble, the transportation will not happen. Visual Control system was set up by empty containers at the main assembly line to set the part transportation, which can be calculated from the number of boxes in the part transportation system resulted in 10 containers from the equation as follows:

$$\text{number of part box in system} = \text{number of parts box at the receiving} + \text{number of parts box at the delivery} + \text{number of part box on milkrun} \quad (2)$$

$$\text{number of parts box at the receiving/delivery} = \frac{Cycle\ time}{Take\ time} + \text{Safety Stock}$$

This also made it possible to determine Visual Control method as shown in Figure 9.

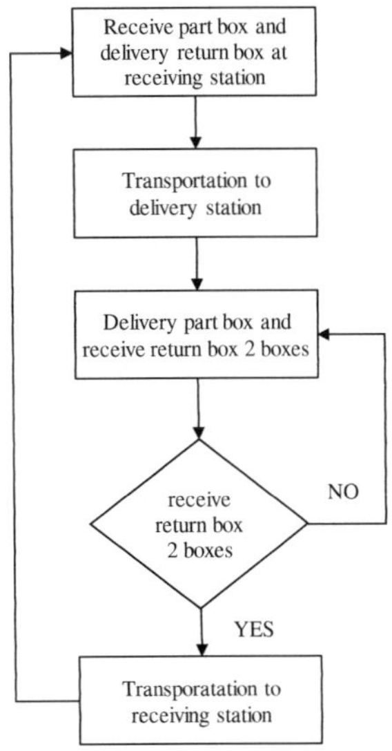

Figure 9. Visual Control Method.

4. Results and discussion

After performing the improvement with ECRS and pull production system as tools to get rid of wastes, the time spent to transport part from a sub-assembly line to the main assembly line dropped from 1,812 seconds to 625 seconds. It was 65.5% working time reduction from the original work as seen in Table 5 and resulted in the need for only 2 electric carts from originally needing 3 carts so it the cost of transportation could be saved by 540.00 baht per year.

Table 5. Flow process chart after improvement.

FLOW PROCESS CHART Page 1 of 1

	Summary			
Location : Agricultural Machinery Assembly Plant				
Activity : Supply Part by Milkrun	**Event**	**Present**	**Proposed**	**Saving**
Date : 2 JULY 2017	Operation	150		
Operator : -	Transport	475		
Circle appropriate Method and Type :	Delay	-		
Method : [Present] Proposed	Inspection	-		
Type: [Worker] Material Machine	Storage	-		
Remark :	Time (second)	625		
	Distance (m)	216		
	Cost	-		

Event Decription	Symbol ○ ⇨ D □ ▽	Time (Second)	Distance (Meter)	Method Recommendation
1. Receive part box and delivery return box of Gear 1, Gear 2 at receiving station.	● ⇨ D □ ▽	20		
2. Transport to receive Brake Ass'y.	○ ➡ D □ ▽	25	4.5	
3. Receive Brake Ass'y.	● ⇨ D □ ▽	20		
4. Transport to receive Hydraulic Ass'y.	○ ➡ D □ ▽	20	3.0	
5. Receive part box and delivery return box of Hydraulic Ass'y at receiving station.	● ⇨ D □ ▽	15		
6. Transport to receive Lever.	○ ➡ D □ ▽	25	4.0	
7. Receive part box and delivery return box of Lever at receiving station.	● ⇨ D □ ▽	10		
8. Transport to receive RR Axle.	○ ➡ D □ ▽	30	4.0	
9. Receive part box and delivery return box of RR Axle at receiving station.	● ⇨ D □ ▽	10		
10. Transport to delivery Gear 1.	○ ➡ D □ ▽	75	60.0	
11. Delivery part box and receive return box of Gear 1 at delivery station.	● ⇨ D □ ▽	10		
12. Transport to delivery Gear 2.	○ ➡ D □ ▽	10	2.0	
13. Delivery part box and receive return box of Gear 2 at delivery station.	● ⇨ D □ ▽	10		
14. Transport to delivery Brake Ass'y.	○ ➡ D □ ▽	60	25.0	
15. Delivery Brake Ass'y at delivery station.	● ⇨ D □ ▽	20		
16. Transport to delivery Hydraulic Ass'y.	○ ➡ D □ ▽	20	4.0	
17. Delivery part box and receive return box of Hydraulic Ass'y at delivery station.	● ⇨ D □ ▽	15		
18. Transport to delivery Lever.	○ ➡ D □ ▽	25	5	
19. Delivery part box of Lever at delivery station.	● ⇨ D □ ▽	5		
20. Transport to delivery RR Axle.	○ ➡ D □ ▽	10	2	
21. Delivery part box of RR Axle at delivery station.	● ⇨ D □ ▽	5		
22. Transport to receive return box of RR Axle.	○ ➡ D □ ▽	5	0.5	
23. Receive return box of RR Axle.	● ⇨ D □ ▽	5		
24. Transport to receive return box of Lever.	○ ➡ D □ ▽	10	7	
25. Receive return box of Lever.	● ⇨ D □ ▽	5		
26. Return to receiving station of Gear 1.	○ ➡ D □ ▽	160	95	

For the implementation of the improvements, the mean efficiency of the transportation of part supply in agricultural machinery assembly plant was about 98 %, which was higher than the original being at approximately 83% as information on part delivery problems shown on Table 1. The improvements has also made the transportation systems to synchronize between parts delivering vehicles and the main assembly line. And all result show as Table 6.

Table 6. Result after improvement.

	Before	After	
Transportation Distance	622	216	meter
Transportation Route	3	1	Route
Transportation Time	1,812	625	second per cycle
Transportation Metod	Push System	Pull System	-
% Line Balance	75.9	79.7	percent
Electric Cart	3	2	car
Manpower	3	2	person
Transportation Efficiency	83	98	percent

5. Conclusion

The objective of this research was to improve the efficiency of the part transportation process caused by delayed transportation and the inability of the main assembly line using the ECRS principle of eliminating waste and pull production systems to synchronize transport system. This could lead to the improvement of the efficiency of systems and processes, transportation to 98 %. This results in the ability to improve the transportation system deliver the parts to meet the customer demand on time and at the desired location.

6. References

[1] Daryl P., Jan R., and Jan OlaS., 2013. Lean production and ERP systems in small-and medium-sized enterprises: ERP support for pull production, *International Journal of Production Research,* **19** , 395-397.

[2] Ziengs, N., Riezebos J., and Germs R., 2012, Placement of effective work-in-progress limits in route-specific unit-based pull systems, *International Journal of Production Research*, **15**, 4359-4362.

[3] Liker, Jeffrey K., 2008, *Toyota culture : the heart and soul of the Toyota way*, New York : McGraw-Hill, 562.

[4] Liker, Jeffrey K., 2006, *The Toyota way fieldbook : a practical guide for implementing Toyota's 4Ps*, New York : McGraw-Hill, 475.

[5] Ana G., 2017, ANALYSIS OF A PRODUCTION SYSTEM WITH THE HELP OF LEAN MANUFACTURING TOOLS, *TEHNOMUS - New Technologies and Products in Machine Manufacturing Technologies*, P88-93.

[6] Bhasin S., 2015, *Lean Management Beyond Manufacturing : A Holistic Approach*, Cham : Springer, 300.

[7] Wickramasinghe V., Implementation of lean production practices and manufacturing performance, *Journal of Manufacturing Technology Management*. 2017, Vol. 28 Issue 4, P531-550.

[8] Jonathan T., and Jeffrey U., The Principles of Lean Manufacturing, *Franklin Business & Law Journal*. 2016, Vol. 2016 Issue 2, P57-70.

[9] Dávid G., András P., Thomas S. and József V., 2013, Milkrun Vehicle Routing Approach for Shop-floor Logistics, *In Forty Sixth CIRP Conference on Manufacturing Systems 2013, Procedia CIRP 2013* 7:127-132.

[10] David H., 1999, *Just in time / David Hutchins*, Hamphsire : Gower, 236.

[11] Lee S., Kang Y. and Prabhu U., Smart logistics: distributed control of green crowdsourced parcel services., *International Journal of Production Research*, Dec2016, Vol. 54 Issue 23, P6956-6968

IOP Conf. Series: Materials Science and Engineering **311** (2018) 012011 doi:10.1088/1757-899X/311/1/012011

Seed defective reduction in automotive Electro-Deposition Coating Process of truck cabin

Aekkalag Sonthilug[1] and Parames Chutima[1,2*]

1 Department of Industrial Engineering, Chulalongkorn University, Bangkok, Thailand 10330
2 Regional Centre for Manufacturing Systems Engineering, Faculty of Engineering Chulalongkorn University, Bangkok, Thailand 10330
E-mail: 1Aekkalag.S@student.chula.ac.th, 2cparames@chula.ac.th

Abstract. The case study company is one of players in Thailand's Automotive Industry who manufacturing truck and bus for both domestic and international market. This research focuses on a product quality problem about seed defects occurred in the Electro-Deposition Coating Process of truck cabin. The 5-phase of Six Sigma methodology including D-Define, M-Measure, A-Analyze, I-Improve, and C-Control is applied to this research to identify root causes of problem for setting new parameters of each significant factor. After the improvement, seed defects in this process is reduced from 9,178 defects per unit to 876 defects per unit (90% improvement)

1. Introduction

The Electro-Deposition Coating (ED) is one of the most important processes which widely used in painting process in automotive industry in order to protect corrosion and rust. This process is used for preparing truck or car surface being smooth before delivery to topcoat process. ED coating process can be divided into 2 sub-processes, namely Pre-treatment process and Electro-Deposition Process. The historical data from October 2016 – March 2017 of the case study company showed that seed defective was found as the highest defect in the ED coating process (84.1% of all defects) which directly impact to the quality of products.

This research applies the Six Sigma methodology for process improvement in order to reduce seed defects. There are 5 phases in the Six Sigma DMAIC process, i.e. 1. Defining Phase – to investigate for getting basic understanding of the process and then specify a problem to solve, 2. Measure Phase – focusing on measurement system analysis to determine the current performance and quantify the problem, 3. Analyze Phase – analyzing by brainstorming and statistic methodology to identify significant factors affecting the problem, 4. Improve Phase – to identify a solution to the problem by designing experiment to find out the optimal key factors, and 5. Control Phase – ensuring that the result from improve phase are well-maintained by creating document procedure in order to control the process and for reacting to any problem in time.

2. Define phase

Define phase is an important first step to identify the problem, target, and scope of the project. In this case study, ED Coating Process of Truck Cabin is selected to study because many of seed defects were found as a major problem in product's quality. Regarding the data of case study company in October 2016 to March 2017 in Figure 1 show that 84.1% of all quality issues comes from seed defect.

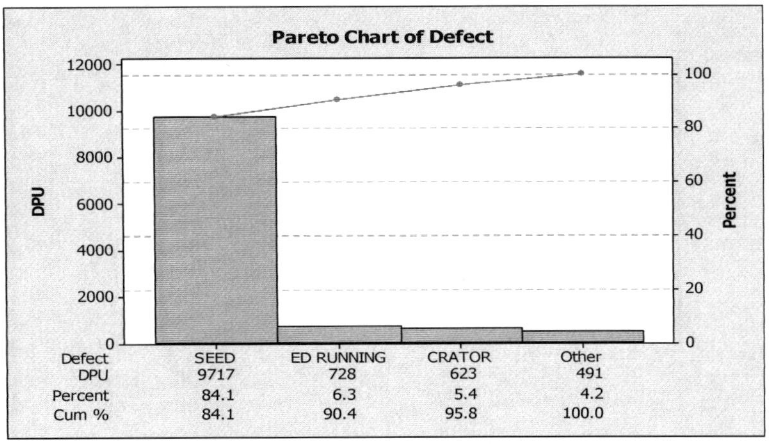

Figure 1. Pareto Chart of defects (Source: Production data of ED coating as of Oct'16-Mar'17).

3. Measure phase

Measure Phase consists of 2 main steps, i.e. Measurement System Analysis (MSA) and Problem Identification Analysis.

3.1 Measurement system analysis (MSA)

In this case study, the quantity of seed defects occurred in ED Coating process of truck cabin is measured for determining quality of the product. Measurement System Analysis (MSA) is used by experimenting 2 operators who perform self-inspection on truck cabin after finished ED coating process. To ensure accuracy and precision of measuring instrument, this approach is repeatedly performed by randomly inspection on 20 truck cabins. The result from the experiment is shown in Table 1 which can be stated that the measurement system is acceptable in term of accuracy and precision

Table 1. Attribute data test performance operator.

Index	Operator A	Operator B	Both operators
%Repeatability	100%	100%	-
%Attribute Score	100%	100%	-
%Reproducibility	-	-	100%
%Attribute Screen Effective Score	-	-	100%

3.2 Problem identification analysis

The expert team brainstorms to list up all potential factors affecting the seed defect in ED Coating process of truck cabin. Next, a Fishbone diagram or Cause-and-Effect matrix is applied in this study as a tool for analyzing and finding out root causes of the problem by categorized relevant factors with 5M1E methodology. It was found that the number of tentative causes is 21 factors. These 21 factors are analyzed the cause relation by scoring these relationships on cause-and-effect matrix. After that, the relevant factors are descending sorted according to Pareto principle (The 80/20 rule). There are 12 factors shown in Figure 2 significantly impacts on seed defects in ED coating process of truck cabin. These factors are further analyzed in the next step by scoring (RPN) each factor by Failure Mode and Effect Analysis (FMEA). The result in Figure 3 show that three highest scored factors, including 1.The proportion of chemicals, 2. Dipping time of degreasing process, and 3. Temperature of chemicals in the degreasing tank, which is 64.7% of the total score are selecting for designing of experiment (DOE) in the next analyze phase.

IOP Conf. Series: Materials Science and Engineering **311** (2018) 012011 doi:10.1088/1757-899X/311/1/012011

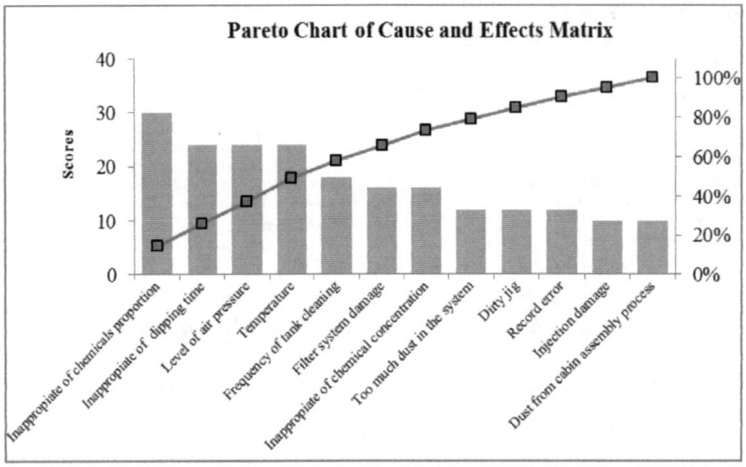

Figure2. Pareto chart of cause and effect Matrix.

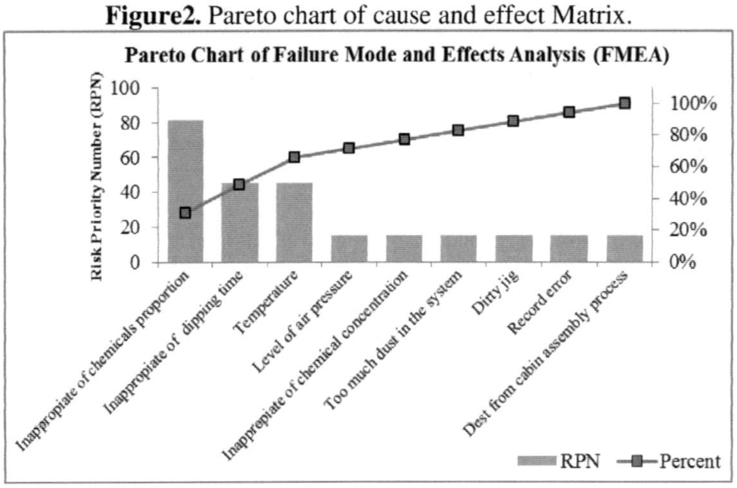

Figure3. Pareto chart of failure mode and effect analysis (FMEA).

4. Analyze phase

In analyze phase, The significant factors, which are screened by Pareto diagram and Failure Mode and Effect Analysis (FMEA), including the proportion of chemicals, dipping time of degreasing process, and temperature of chemicals in the degreasing tank are selected to conduct the experimental design method by adopting the Box-Behnken to analyze the root cause of the defect. The experiment design conducts 3 factors and 3 levels each as shown in Table 2. The response variable of this experimental is coating weight which measuring the result by x-ray fluorescence (XRF) Model RIGAKU, RIX-2000. Regarding the standard of case study company, the coating weight required to equal 3-5 g/ m^2.

Table 2. Factors level in Box-Behnken design in degreasing process.

No.	Input factors	Symbols	Unit	Factors Level		
				Min (-1)	Med (0)	Max (+1)
1	Chemicals proportion	Chemical	%	25	62.5	100
2	Dipping time	Time	Seconds	180	210	240
3	Temperature	Temp	Degrees Celsius	45	50	55

IOP Conf. Series: Materials Science and Engineering **311** (2018) 012011 doi:10.1088/1757-899X/311/1/012011

The result from the experiments are statistically analyzed by Minitab software as Figure 4 shows that the main effect Proportion, Time, and Temperature are significant at 95% confident interval. In addition, two-way interaction of Chemical proportion*Temperature and Temperature*Time are also significant.

Response Surface Regression: Result versus % New Chemical, Temp, Time

Analysis of Variance for Result

Source	DF	Seq SS	Adj SS	Adj MS	F	P
Regression	9	6.95271	6.95271	0.772523	173.28	0.000
Linear	3	6.50188	0.29439	0.098131	22.01	0.003
%New Chemical	1	6.21281	0.05484	0.054841	12.30	0.017
Temp	1	0.06125	0.15319	0.153190	34.36	0.002
Time	1	0.22781	0.16788	0.167885	37.66	0.002
Square	3	0.26708	0.26708	0.089028	19.97	0.003
%New Chemical*%New Chemical	1	0.08103	0.10776	0.107756	24.17	0.004
Temp*Temp	1	0.07830	0.09256	0.092564	20.76	0.006
Time*Time	1	0.10776	0.10776	0.107756	24.17	0.004
Interaction	3	0.18375	0.18375	0.061250	13.74	0.008
%New Chemical*Temp	1	0.10563	0.10563	0.105625	23.69	0.005
%New Chemical*Time	1	0.00250	0.00250	0.002500	0.56	0.488
Temp*Time	1	0.07563	0.07563	0.075625	16.96	0.009
Residual Error	5	0.02229	0.02229	0.004458		
Lack-of-Fit	3	0.01062	0.01062	0.003542	0.61	0.671
Pure Error	2	0.01167	0.01167	0.005833		
Total	14	6.97500				

S = 0.0667708 PRESS = 0.19625
R-Sq = 99.68% R-Sq(pred) = 97.19% R-Sq(adj) = 99.11%

Figure 4. The result of ANOVA Box-Behnken design in degreasing process.

5. Improvement phase

All of 3 significant factors derived from previous phase including proportion, dipping time, and temperature need further analysis in improvement phase to find the optimum conditions in order to reduce seed defect in Electro-Deposition Coating Process of Truck Cabin by using full factorial design in the experiment. The experiments were randomly run with 2 replicates. The result of the experiments analyzed by the Minitab software as shown in Figure 5 demonstrates that the optimum condition of each factor should be the proportion of new chemical is 100%, temperature is 45 degrees Celsius, and dipping time is 180 seconds.

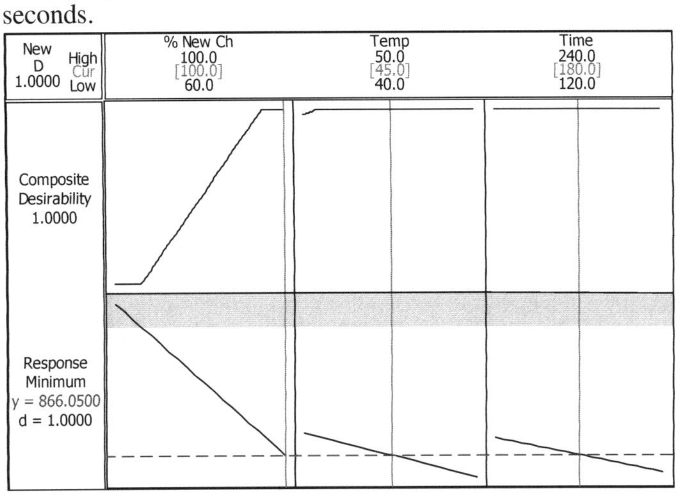

Figure 5. New Parameter Setting for degreasing process.

6. Control phase

According to the result from improvement phase, the new parameters setting of the significant factors are implemented in the current process and the defect data in the Electro-Deposition Coating Process of Truck Cabin are collected for 30 days. It proves that the quantity of seed defects is significantly reduced as shown in Figure 6.

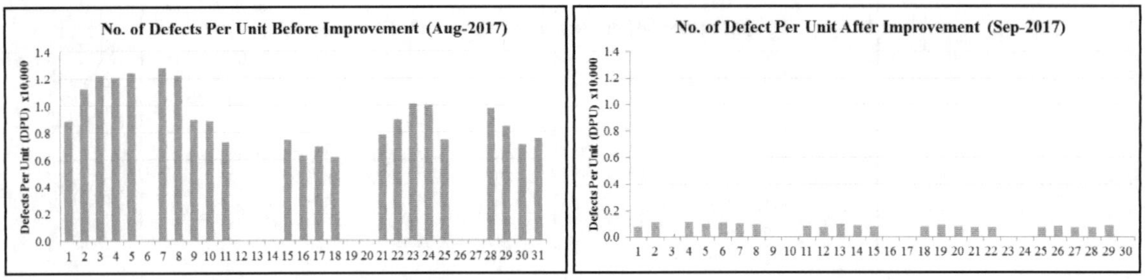

Figure 6. Comparison number of defects per unit in production line.

To maintain the quality of the process after the improvement, Work instruction and control plan of this process are created. Moreover, an operator training is essential for clear understanding the importance of parameter controlling to ensure that the performance is maintained at this good level.

7. Conclusion

The objective of this research is for reducing seed defects of truck cabin in the Electro-Deposition Coating Process by applying the Six Sigma methodology to improve the process. This research study follows the DMAIC steps of the Six Sigma methodology to find the root causes that affect to seed defects on truck cabin and determine the optimum conditions for setting set as new parameters. The result of the experiment shows that the quantity of seed defects of truck cabin in the Electro-Deposition Coating Process is reduced from 9,178 defects per unit to 876 defects per unit (90% improvement).Work instruction and training are created to maintain this good performance.

8. References

[1] Pyzdek T and Keller P A 2010 *The Six Sigma Handbook, A Complete Guide for Green Belts, Black Belts, and Manager at All levels* (3rd ed) (New York: McGraw-Hill) pp 147-465

[2] Park S H, Lee M J and Chung M Y 1999 *Theory and Practice of Six Sigma* (Seoul: Korean Standards Association) p 210

[3] Bertels T 2003 *Rath & Strong's Six Sigma Leadership handbook* (USA: John Wiley &Sons) pp 196 - 260

[4] Harry M and Schroeder R 2000 *Six Sigma: The Breakthrough Management Strategy Revolutionizing the World's Top Corporations* (New York: Doubleday) pp 138-230

[5] Michael L and David R 2005 *Lean Six Sigma Pocket Tool Book* (USA: The McGraw-Hill) pp 131-196

[6] Krogstie L and Martinsen K 2013 *Proc. 46th CIRP Conf. on Manufacturing System* (Amsterdam: Elsevier) vol 7 pp 610 – 615

[7] Pugna, A, Negrea R and Miclea S 2016 *Proc. Int. Symp. on Social and Behavioral Sciences* vol 221 (Amsterdam: Elsevier) pp 308 – 316

[8] Douglas C and George C 2003 *Applied Statistics and Probability for Engineers* (3rd ed) (USA: John Wiley & Sons) pp 468-571

[9] Jack P Pekar 1995 *Total Quality Management: Guiding Principles for Application* (USA: American Society For Testing Material) pp 47-54

[10] Douglas C Montgomery 2005 *Design and Analysis of Experiments* (USA: John Wiley & Sons) pp 235-283

IOP Conf. Series: Materials Science and Engineering **311** (2018) 012012 doi:10.1088/1757-899X/311/1/012012

Springback law study and application in incremental bending process

Feifei Zhang[1], Kai He[1], Xiaobing Dang[2] and Ruxu Du[2]

1. Shenzhen Institute of Advanced Technology, Chinese Academy of Sciences, Shenzhen, China
2. Institute of Precision Engineering, The Chinese University of Hong Kong, Hong Kong

E-mail: ff.zhang@siat.ac.cn

Abstract. One incremental bending process has been proposed for manufacturing the complex and thick ship-hull plates. The accuracy and efficiency for this novel process is mainly dependent on the loading path and thus the unavoidable springback behavior should be considered in the loading path determination. In this paper, firstly, the numerical simulation method is verified by the corresponding experiment, and then the springback law during the incremental bending process is investigated based on numerical simulation, and later the loading path based on the springback law and the minimum energy method is achieved for specific machining shape. Comparison between the designed curve based on springback law and the new simulation results verifies that the springback law obtained by numerical simulation is believable, so this study provides a new perspective for the further research on incremental bending process.

1. Introduction

Recently, one novel incremental bending process has been proposed for manufacturing the thick plates with much smaller loading force [1]. One of the most important issues to guarantee a good part with this method is to have a proper loading path. In our former study, minimum energy method and Euler-Bernoulli beam theory [2] were used to achieve the loading path. However, the whole incremental bending process is finished step by step while after each step there is huge springback phenomenon caused by stress relief. The loading path obtained based on minimum energy method and Euler-Bernoulli beam theory without considering springback would induce huge error in the final shape of the deformed part. So springback after each bending process should be considered carefully.

In recent years, many scholars have done researches on the factors that influence the springback phenomenon during bending process. For example, Ozgu et al. (2014) [3] conducted bending experiments for some steel materials and they found springback reduces with large bending angle. Michael et al. (2016) [4] investigated the springback behavior of two steels based on free bending experiments. Results show that bigger punch velocity can reduce springback while flat length has little influence on springback. Wiriyakorn and Sutasn (2017) [5] discussed the influence of channel width on springback characteristics in the U-bending process and they found that for the case without pressure pad, bigger channel width could induce bigger springback. Besides, Botros (1967) [6] proved that springback increases as the thickness of the blank decreases. Ma (2015) [7] verified that grain size could influence the stress distribution and thus effect the springback angle.

Review of the previous researches showed that many experimental conditions can influence the springback behavior, so efforts can be made to improve the springback prediction. However, most of

the research just qualitatively analysed the changing regularity of springback behavior. For the novel incremental bending process, the quantitative relationship between the springback amount and the initial bending shape should better be obtained to get the proper loading path. So far, some researchers have done such studies. For example, Gardiner (1957) [8] developed one generalized mathematical formula to describe the relationship of the final bending radius and the initial bending radius. But this analysis can only be used to predict the springback of the elastic-perfectly plastic materials. Hosford and Caddell [9] proposed another classic springback model for strain hardening materials. Besides, Zhang et al. (2007) [10] calculated the angular change during springback for U-bending process. The disadvantage of those springback models mentioned above is that it is hard to get the accurate shape directly after each springback process. Some new and complicated constitutive models have also been proposed to predict the springback behavior accurately [11, 12] while they are impossible to obtain the springback amount directly. Wang et al. (2008) [13] obtained the relationship between the unload angle and punch displacement for air bending process when the punch stamps only at the mid-span of the sheet metals. Their study provided a good air bending angle control system while the springback behavior is still unknown when the punch stamps at other positions of the sheet metals.
Generally speaking, it is really difficult to obtain an intuitive theoretical model to describe the springback behavior in our incremental bending process. So in this paper, the numerical simulation method is used instead to find the springback law in the incremental bending process and to guide the loading path determination.

2. Basic theory of incremental bending process

2.1 Incremental bending process
Incremental bending process is one gradual forming process for manufacturing complex and thick ship-hull plates. Figure1 (a) shows the schematic diagram of the incremental bending process. During the novel process, the plate is firstly supported by some rotatable and movable supports. Then the punch (or punches) moves down to press the blank. The process may take several steps and during each step the stamping depth and position may different according to the objective part. The process continues until the final shape is achieved. It should be noted that since the boundary of the plate is free, the forming force is significantly reduced compared with the traditional stamping process, which is one obvious advantage of this novel process. So far our group has designed one testing machine to realize the incremental bending process and the experimental setup is shown in Figure 1 (b).

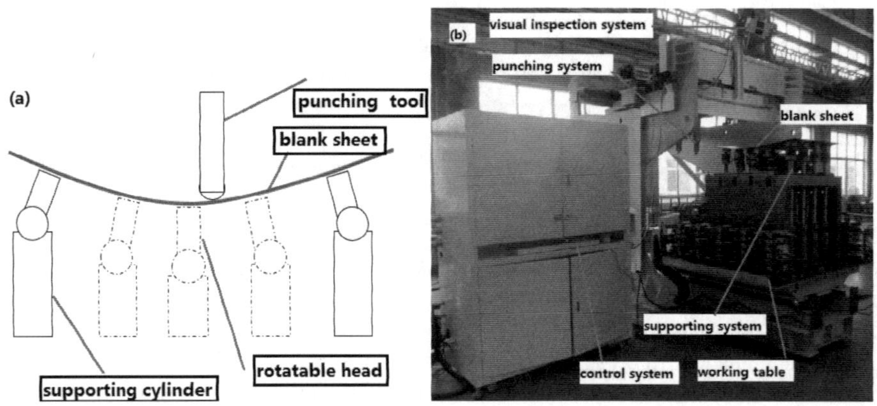

Figure 1. Incremental bending process (a) schematic diagram (b) testing machine.

2.2 Determination of the proper loading path
In order to obtain a ship-hull plate with good quality and high efficiency, the proper stamping loading path should be used, otherwise the bending process may take a much longer time or the deformed shape is not accurate as we expected. Minimum energy method could be used as one principle to

design our loading path. The detailed explanation of the minimum energy method can be found in our published paper [14]. According to minimum energy method, the stamping point should be at where the error between the current shape and objective shape is maximum.

It should be noted that, after each bending process, there is huge inevitable springback phenomenon caused by relief of residual stress in sheet metal forming process. So the error mentioned in minimum energy method should be the one between the objective shape and the shape after each springback process. So how to get the shape correctly after each springback process is very essential for the loading path determination, and that is the key research topic in this study.

3. Verification of numerical simulation analysis

Since it is difficult to get the theoretical model directly for this incremental bending process, the numerical method is used instead to investigate the springback law, so the first issue here is to verify the reliability of the numerical simulation method.

The basic numerical simulation model is shown in Figure 2(a). During the incremental bending process, eight supports can be used to support the plate, and the three punches can move along y-axis and z-axis to make the plate deform. The corresponding experiment was also carried out on our testing machine and the final variable curved plate is shown in Figure 2(b).

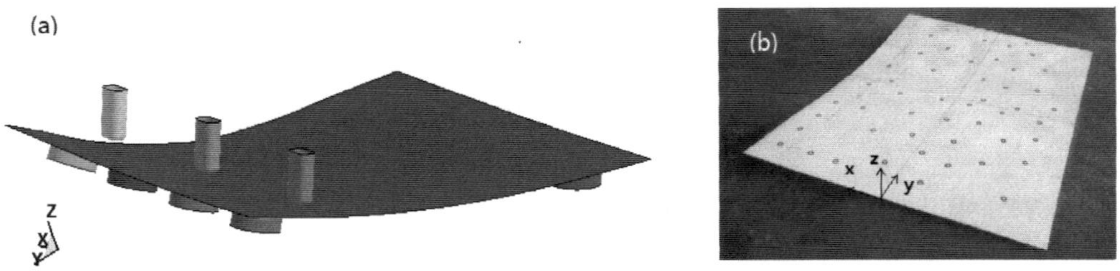

Figure 2. FEM and experimental Models (a) FEM model; (b) experimental result.

Figure 3 shows the bending shape along y-axis of the deformed plate. It can be found that the predicted z-axis displacement of the plate coincides well with the experimental data, which validates the numerical simulation model in the application of this incremental bending process. The maximum difference between the experimental data and numerical ones are about 10mm. Compared to the dimension of the plate (1000mm*800mm*5mm), the numerical results are quite acceptable. The errors are considered to be caused by neglecting the Bauschinger effect and anisotropy of the steel plate. Considering the good results shown in Figure 3, the numerical model used in this study can help us analyze the springback behavior during the incremental bending process.

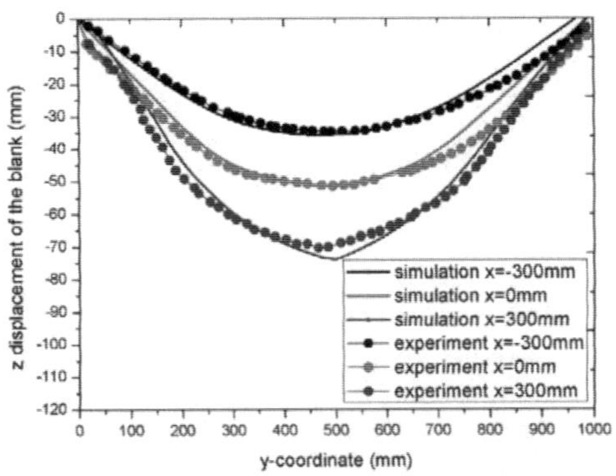

Figure 3. Comparison of experimental and simulation results at different positions.

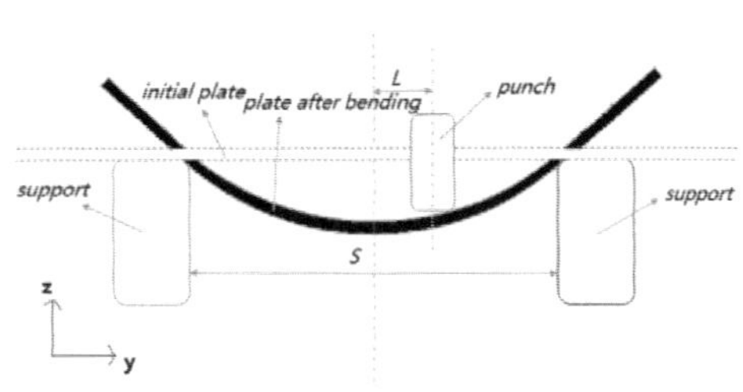

Figure 4. numerical model for springback study.

4. Springback law analysis based on numerical simulation

Here in this section, the numerical simulation method is used to analysis the springback behavior in incremental bending process. The basic numerical model is shown in Figure 4, where L is the distance from punch center to the middle of the two supports, and S is the spacing between two supports. We will change the value of L and S separately to investigate how the springback changes. It should be noted that during one simulation, the punch only moves up and down along z direction to realize different bending depths while it doesn't change along the length direction of the blank (or y-axis).

4.1 Springback study for the case with constant L=0mm

In this section, the springback of the plate is investigated for the situation that the spacing of the two supports changes while the punch always stamps at the middle position of the supports (that is, L=0mm).

Figure 5 shows the geometry of the plate after each bending process and after each springback process, where 'first-before spbk' means the shape of the blank after the first bending process, 'first-after spbk' means the shape of the blank after first springback process. From the results, we can get some interesting phenomenon.

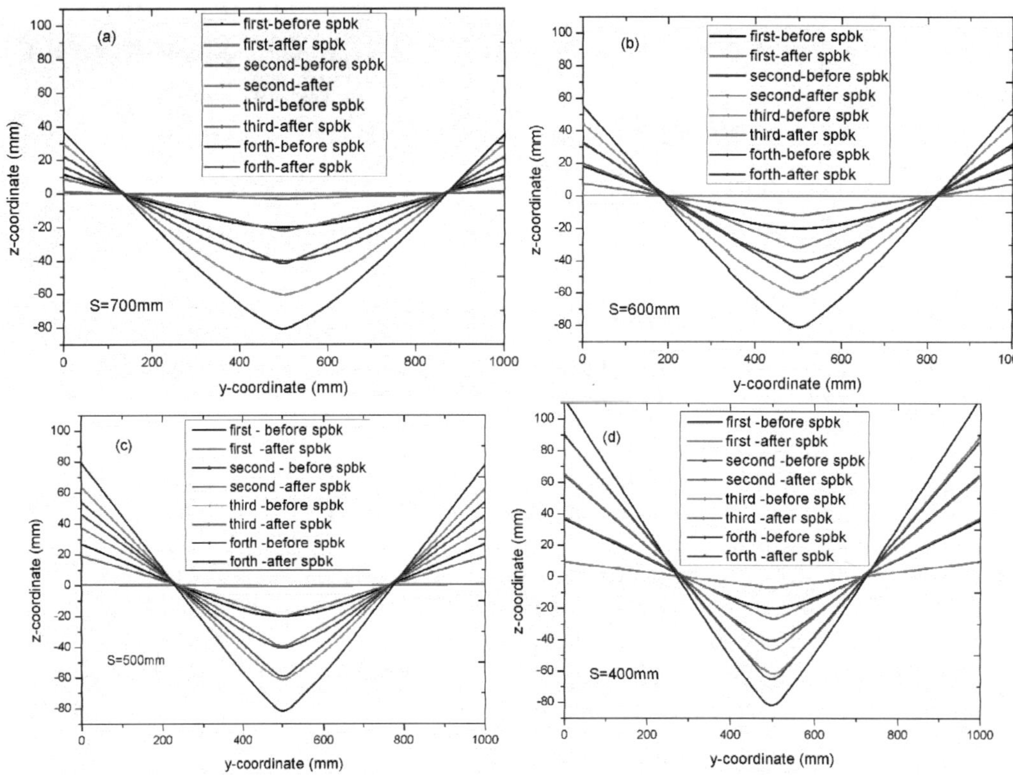

Figure 5. The geometry of the plate before springback and after springback for various bending steps for L=0 mm.

First, results show that when the stamping depth is smaller, the geometry of the plate after bending process is a smooth curve. However, when the stamping depth is larger (bigger than 50mm), the geometry of the plate is more inclined to two intersecting straight lines. This phenomenon can be explained as follows: When the stamping depth is smaller, the deformation of the stamping position of the blank is mostly elastic. According to classic Euler-Bernoulli Beam theory, the deflection for the beam can be described by a simple cubic equation. However, when the stamping depth is larger, the deformation of the stamping position of the blank is mostly plastic, so the deformation is only concentrated on the loading area.

Second, when the spacing of the two supports are smaller, the geometry of the plate after each bending process is more inclined to two intersecting straight lines, which is caused by the truth that the deformation of the stamping position of the blank is more prone to plastic.

Third, the geometry of the plate after each springback process is two intersecting straight lines. That's because after springback, the elastic deformation of the plate is recovered and only the plastic deformation is left. So only the stamping point (or area, depends on the shape and dimension of the punch) has obvious geometry change.

Generally speaking, it is not easy to describe the geometry of the plate after each bending process and each springback process using the same equation, so in the following study, we try to figure out the springback distribution law after each stamping step.

Figure 6 shows the springback amount of the plate for various bending steps. It is clear that the springback curve for the first bending stage in Figure 6(a) (or Figure 6(b)) is different from other cases. From Figure 5(a) (or Figure 5(b)) we can see that the geometry of the plate (shown in red in Figure 5(a) or Figure 5(b)) is one horizontal line after springback, that means during the first stamping process, only elastic deformation occurs in the blank, so the springback curve for the first bending stage is equal to the stamping curve. For other stamping steps, there is plastic deformation, and results show that all the other springback curves after different stamping depths are close to each other when the spacing of the supports are constant. Besides, when the spacing of the supports are smaller, the maximum springback amount is smaller due to the bigger plastic deformation.

According to the observation from Figure 5 and Figure 6, we find that though it is difficult to describe the whole stamping process and springback process with the same equation, the maximum springback amount (the springback at the stamping position) only depends on the spacing of the supports if there is plastic deformation during the stamping process. So the geometry of the plate after each springback step can be predicted only if the maximum springback amount is given.

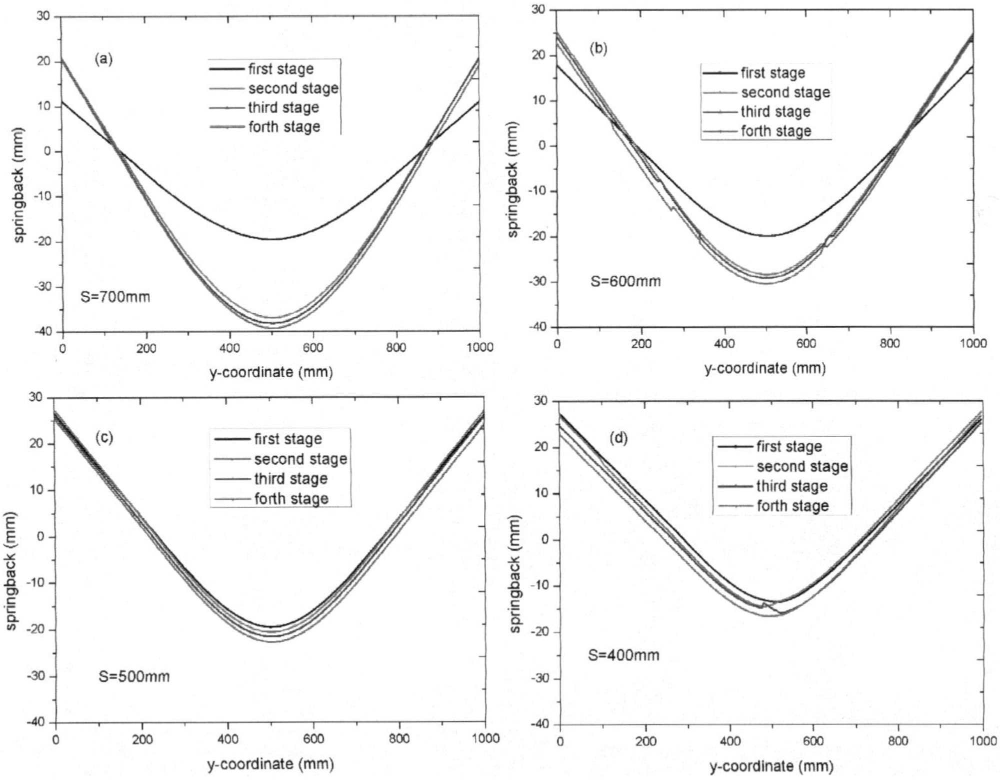

Figure 6. The springback amount of the plate for various bending steps for L=0 mm.

4.2 Springback study for the case with constant S=700mm

In the former section, the stamping point is only at the mid-span of the blank, here in this section, the springback of the plate is investigated for the situation that the spacing of the two supports is constant (S=700mm) while the punch stamps at different positions between the supports.

Figure 7 shows the geometry of the plate after each bending process and after each springback process, and Figure 8 shows the springback amount of the plate for various bending steps. Similar to the phenomenon shown in Figure 5 and Figure 6, the maximum deformation occurs in the stamping point. Besides, since the springback is elastic process, the springback curves shown in Figure 8 satisfy the deflection equation of Euler-Bernoulli beam theory, which can be described by a simple cubic equation.

For each sub-figure of Figure 8, the horizontal distance of the stamping point to the support is the same (that is L=constant) and the springback amount for each bending steps are close to each other, so we can roughly regard that the springback amount is constant no matter how much the bending depth is. However, it is clear from Figure 8 that when the stamping point is more close to the support, the springback amount of the stamping point is smaller.

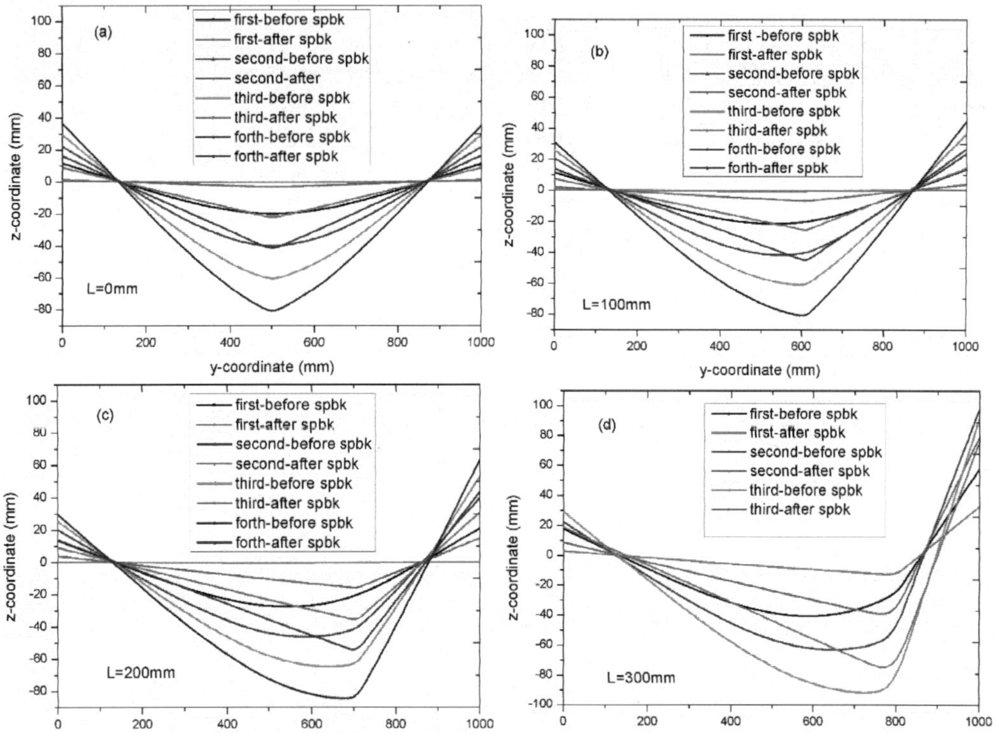

Figure 7. the geometry of the plate before springback and after springback for various bending steps for S=700 mm.

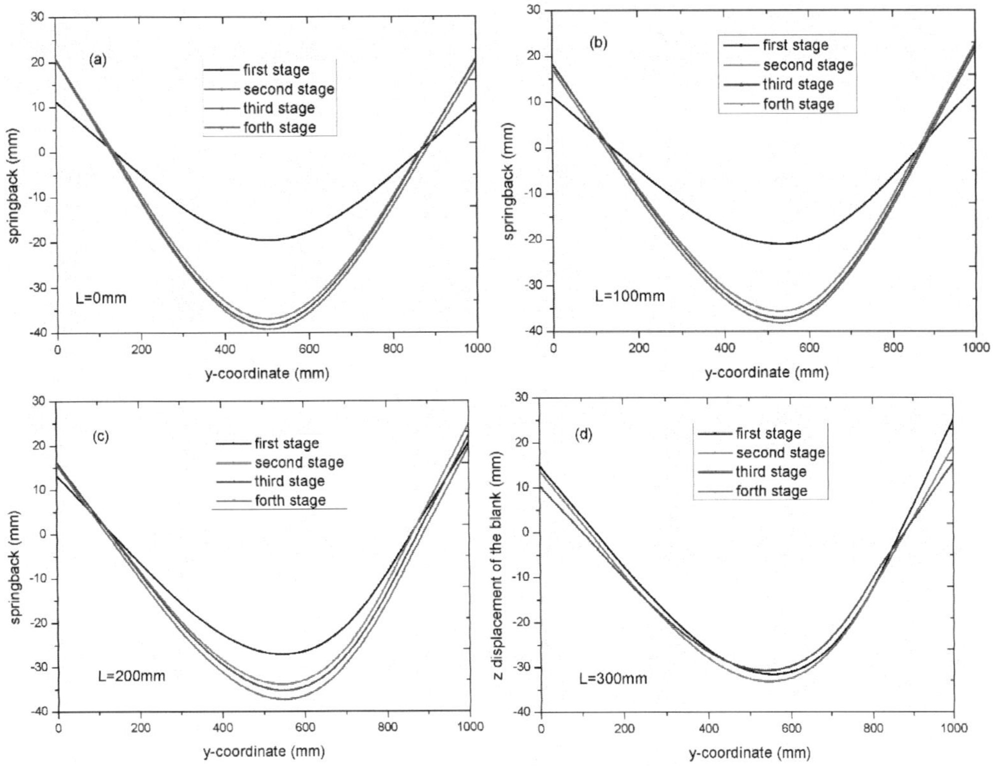

Figure 8. The springback amount of the plate for various bending steps for S=700 mm.

4.3 Springback law summary

According to Figure 5- Figure 8 we can summary the springback law for the cases with different S and L. Figure 9 (a) shows the springback amount of the stamping point in the middle point of two supports when the spacing of the supports changes. According to the distribution disciplinarian of the springback amount, a quadratic equation can be used to fit the data and obtain other springback amount for different spacing of the supports (or S). Figure 9 (b) shows the springback amount of the stamping point at different positions when the spacing of the supports is 700 mm. Generally speaking, we can regard that the springback amount of the middle point varies linearly with the distance from the middle point to the punch stamping point. So based on the data of Figure 9, the springback amount for different positions when the spacing of two supports are different can be obtained, which will help to obtain the shape of the plate after each springback process and then achieve the proper loading path.

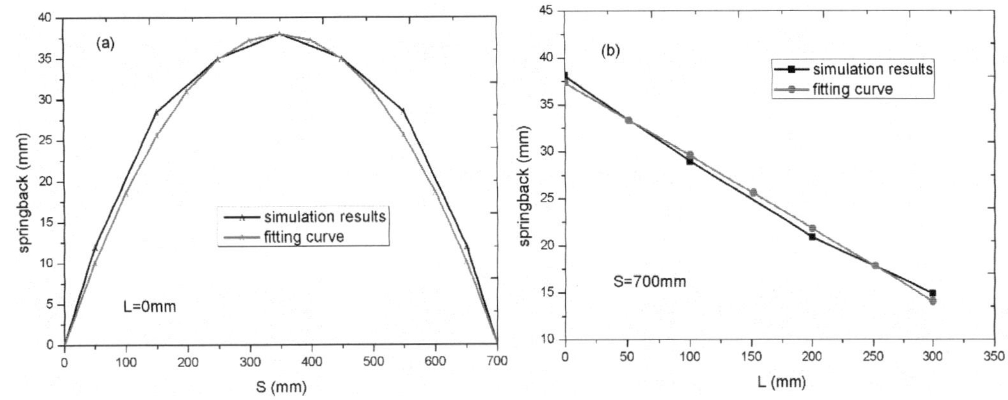

Figure 9. Springback law (a) for the case with S=700 mm (b) for different gaps.

5. Application of the springback law in loading path decision

Here in this section, the proper loading path of the incremental bending process can be obtained based on minimum energy method and the above springback law. Taking the objective shape y=x^2*100/350^2-100 as an example, the green line in Figure 10 shows the objective shape, while the blue star points show the stamping positions and the blue curves show the achieved shapes of the blank after each stamping and springback process. Numerical simulation with the same loading path is also conducted. Actually, each step during the incremental bending process could be compared between the new simulation result and the designed curve based on the springback law. Figure 11 shows the comparison of some step between simulation results and the designed curve. Results show the new simulation results coincide well with the calculation results. The small error between them could be caused by neglecting the tool shape and the slight difference shown in each sub-figure for Figure 6 and Figure 8. However, the error shown in Figure 11 is quite acceptable, so the loading path determination method based on the springback law is validated.

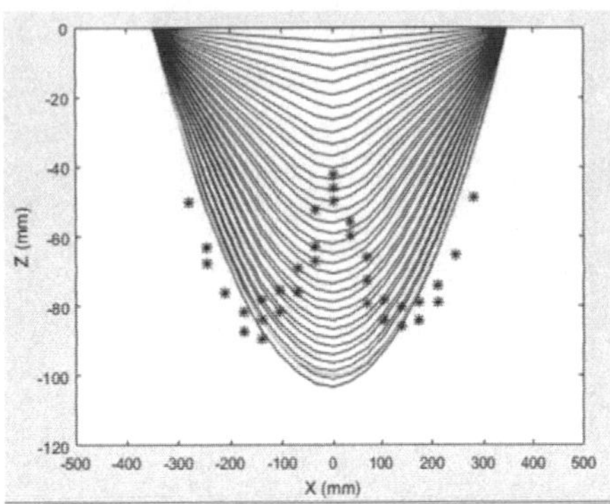

Figure 10. the loading path based on minimum energy method and springback law.

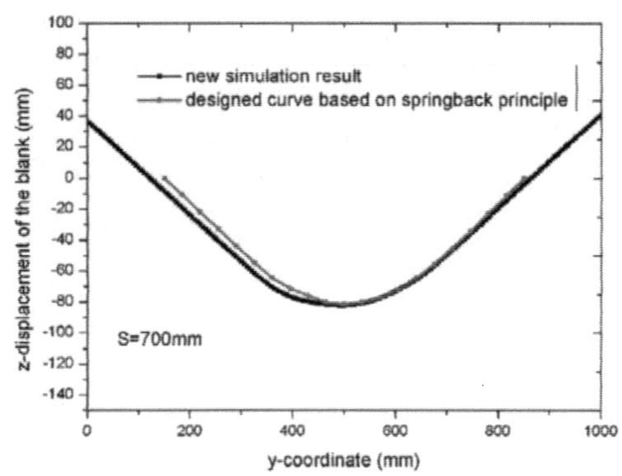

Figure 11. the comparison of calculation curve and simulation result.

6. Conclusions

In incremental bending process, one important issue is to achieve a proper loading path. The unavoidable springback behaviour should be considered in the loading path determination. Here in this paper, the springback law is investigated based on numerical simulation, and the loading path based on the springback law and minimum energy method is used to guide the incremental bending process. Some key conclusions are summarized as follows:

- The maximum bending amount after each bending process is near the stamping position when there is plastic deformation.
- For the case that the spacing of the supports are constant, the springback amount for the same stamping position is the same no matter how many bending processes have been conducted In other words, the springback amount for the same stamping position is independent with the loading path.
- When stamping the middle point of the two supports, the springback amount for the stamping position can be expressed as a quadratic function of the stamping point position.
- When the spacing of the two supports is constant, the springback amount for the stamping position varies linearly with the spacing from the middle point to the punch stamping point.

7. References

[1] Li J, Dang X, He K, Zuo Q and Du R 2016 *T. Can. Soc. Mech. Eng.* **40** pp 433-443
[2] Bauchau O A and Craig J I 2009 *Structure Analysis* (Springer) pp 173
[3] Senol O, Esat V and Darendeliler H 2014 *Procedia Eng.* **81** pp 999-1004
[4] Krinninger M, Opritescu D, Golle R and Volk W 2016 *Procedia Cirp.* **41** pp 1066-71.

[5] Wiriyakorn Phanitwong, Sutasn Thipprakmas. *Procedia Eng.* **183** pp 17-22
[6] Botros B MS 1967 *ASME WA/PROD* **17** pp 67
[7] Ma Z, Tong G, Chen F, Wang Q and Wang SC 2015 *J. Mater. Process. Tech.* **224** pp 11-17.
[8] Gardiner FJ. The springback of metals. *Transactions of ASME* 1957, 79:1-9
[9] Hosford W, Caddell R. 2007 *Metal forming –mechanics and metallurgy.* (New York: Cambridge University Press) p 241
[10] Zhang D, Cui Z, Ruan X and Li Y 2007 *Com. Mater. Sci.* **38** pp 707-715
[11] Badr O M, Rolfe B, Zhang P and Weiss M 2017 Int. J. Mech. Sci. **128-129** pp 389-400
[12] Yang X, Choi C, Sever N K and Altan T 2016 *Int. J. Mech. Sci.* **105** pp 266-272
[13] Wang J, Verma S, Alexander R and Gau JT 2008 *J. Manuf. Process.* **10** pp 21-27
[14] Dang X, He K, Li W, Zuo Q and Du R 2017 *J. Manuf. Sci. Eng.* **139** 071009

Acknowledgement
The research is supported by Shenzhen Basic Research Project (JCYJ20150521094519483) and the Key Cooperative Program of the Bureau of International Cooperation, Chinese Academy of Sciences (172644KYSB20160024) and Chinese Postdoctoral Science Foundation (2017M612779).

IOP Conf. Series: Materials Science and Engineering **311** (2018) 012013 doi:10.1088/1757-899X/311/1/012013

Ideal flow theory for the double – shearing model as a basis for metal forming design

S Alexandrov[1,2] and N T Trung[3,4]

1 Visiting Professor, Division of Computational Mathematics and Engineering, Institute for Computational Science, Ton Duc Thang University, Ho Chi Minh City, Vietnam
2 Visiting Professor, Faculty of Civil Engineering, Ton Duc Thang University, Ho Chi Minh City, Vietnam
3 Professor, Division of Computational Mathematics and Engineering, Institute for Computational Science, Ton Duc Thang University, Ho Chi Minh City, Vietnam
4 Professor, Faculty of Civil Engineering, Ton Duc Thang University, Ho Chi Minh City, Vietnam
E-mail: sergeialexandrov@tdt.edu.vn

Abstract. In the case of Tresca' solids (i.e. solids obeying the Tresca yield criterion and its associated flow rule) ideal flows have been defined elsewhere as solenoidal smooth deformations in which an eigenvector field associated everywhere with the greatest principal stress (and strain rate) is fixed in the material. Under such conditions all material elements undergo paths of minimum plastic work, a condition which is often advantageous for metal forming processes. Therefore, the ideal flow theory is used as the basis of a procedure for the preliminary design of such processes. The present paper extends the theory of stationary planar ideal flow to pressure dependent materials obeying the double shearing model and the double slip and rotation model. It is shown that the original problem of plasticity reduces to a purely geometric problem. The corresponding system of equations is hyperbolic. The characteristic relations are integrated in elementary functions. In regions where one family of characteristics is straight, mapping between the principal lines and Cartesian coordinates is determined by linear ordinary differential equations. An illustrative example is provided.

1. Introduction

Ideal plastic flows are those for which all material elements follow minimum work paths [1]. This theory has been fully developed for the constitutive equations comprising the Tresca yield criterion and its associated flow rule [2]. A comprehensive overview of the ideal flow theory has been provided in [3]. Recently, the existence of stationary planar ideal flow solutions has been proven for models of anisotropic and pressure – dependent plasticity [4, 5]. In particular, the model of anisotropic plasticity proposed in [6] has been adopted in [4] and the double slip and rotation model of pressure – dependent plasticity proposed in [7] has been adopted in [5]. The ideal flow theory is used as the basis of procedures for the direct preliminary design of forming processes. A number of such design solutions have been developed in [8 - 13]. In the case of stationary ideal flows, the distribution of plastic strain in the final product is uniform. Moreover, the distribution of the damage parameter is uniform for certain damage evolution laws [14, 15]. The present paper develops a method for constructing a class of stationary planar ideal solutions for the double – shearing model of pressure – dependent plasticity proposed in [16]. This model has been originally proposed for soils. However, it has been shown (see, for example, [17]) that it is adequate for some metallic materials as well.

2. Ideal flow conditions

In the case of the constitutive equations comprising the Tresca yield criterion and its associated flow rule, the principal stress and strain rate directions coincide. The ideal flow condition demands that the trajectories of the greatest (algebraically) principal stress (and strain rate) are fixed in the material [2]. In the case of plane strain deformation, the trajectories of both principal stresses (and strain rates) in planes of flow are fixed in the material. Therefore, the principal lines coordinate system is Lagrangian and solving the boundary value problem of plasticity reduces to a purely geometric problem [18]. In the case of stationary flows, the aforementioned general ideal flow condition requires that the trajectories of the greatest principal stress coincide with streamlines. Solving plane strain problems also reduces to a purely geometric problem [19]. A proof of the existence of stationary bulk planar ideal flows for the double slip and rotation model proposed in [7] has been given in [5]. However, no method for finding solutions has been developed. The model proposed in [7] is based on the Mohr - Coulomb yield criterion and a non-associated flow rule. The principal stress and strain rate directions do not coincide in this model. It has been shown in [5] that the ideal flow condition is that the trajectories of one of the principal stresses in planes of flow coincide with streamlines. It is evident that the ideal flow condition is an additional equation to the standard system of equations of this or that theory of plasticity. However, the proofs of the existence of ideal flows [2, 5, 20] demonstrate that the system of equations comprising the standard constitutive equations and the ideal flow condition is compatible. The difference between the ideal flow solutions and the standard solutions in plasticity lies in boundary conditions. In particular, not all of standard boundary conditions are compatible with the ideal flow condition. In the case of stationary flows, tool surfaces should be frictionless. Moreover, the shape of tool is not fully arbitrary and the design problem is to find such a shape that an ideal flow solution exists.

3. Material model

The double shearing model has been proposed in [16] for soils. Later, it has been shown that this model can be used for metallic materials as well [17]. Under plane strain conditions, the constitutive equations of the model are the Mohr - Coulomb yield criterion and the flow rule. In an arbitrary orthogonal coordinate system (ξ,η) these equations can be written as

$$\left(\sigma_{\xi\xi}+\sigma_{\eta\eta}\right)\sin\phi+\sqrt{\left(\sigma_{\xi\xi}-\sigma_{\eta\eta}\right)^2+4\sigma_{\xi\eta}^2}=2k\cos\phi \tag{1}$$

and

$$\varepsilon_{\xi\xi}+\varepsilon_{\eta\eta}=0,\quad \sin 2\psi\left(\varepsilon_{\xi\xi}-\varepsilon_{\eta\eta}\right)-2\cos 2\psi\varepsilon_{\xi\eta}-2\sin\phi\left(\omega_{\xi\eta}+d\psi/dt\right)=0. \tag{2}$$

Here $\sigma_{\xi\xi}$, $\sigma_{\eta\eta}$ and $\sigma_{\xi\eta}$ are the components of the stress tensor referred to the (ξ,η) coordinate system, $\varepsilon_{\xi\xi}$, $\varepsilon_{\eta\eta}$ and $\varepsilon_{\xi\eta}$ are the components of the strain rate tensor referred to the (ξ,η) coordinate system, $\omega_{\xi\eta}$ is the only non-zero spin (vorticity) component referred to the (ξ,η) coordinate system, ψ is the angle between the $\xi-$ direction and the greatest principal stress measured from the $\xi-$ direction anti-clockwise, d/dt denotes the convected derivative, k is the cohesion, and ϕ is the angle of internal friction. The existence of stationary bulk planar ideal flows has been proven in [5] for the double slip and rotation model proposed in [7]. It is possible to show that this proof is valid for the double shearing model as well. The constitutive equations of the double slip and rotation model are the Mohr - Coulomb yield criterion (1) and the flow rule in the form

$$\varepsilon_{\xi\xi}+\varepsilon_{\eta\eta}=0,\quad \sin 2\psi\left(\varepsilon_{\xi\xi}-\varepsilon_{\eta\eta}\right)-2\cos 2\psi\varepsilon_{\xi\eta}-2\sin\phi\left(\omega_{\xi\eta}+\Omega\right)=0. \tag{3}$$

Here Ω is the intrinsic spin due to grain rotation. Stationary planar ideal flows for the double slip and rotation model exist if $\Omega=0$ [5]. Consider the second equation in (2) and choose the (ξ,η)

coordinate systems such that its $\xi -$ coordinate curves are everywhere coincident with streamlines. The ideal flow condition is that the trajectories of one of the principal stresses in planes of flow are everywhere coincident with streamlines. Therefore, $\psi = 0$ or $\psi = \pi/2$ everywhere. In either case, $d\psi/dt = 0$ everywhere. Therefore, the second equation in (2) coincides with the second equation in (3) at $\Omega = 0$ and the models coincide under the ideal flow condition. Hence, the solution given below is valid for both models.

4. System of equations

Let h_ξ and h_η be the scale factors for the $\xi -$ and $\eta -$ coordinate curves, respectively. The coordinate curves of this coordinate system coincide with trajectories of the principal stresses, σ_ξ and σ_η. Assuming that $\sigma_\xi > \sigma_\eta$ it has been shown in [21] that the scale factors satisfy the equation

$$h_\xi h_\eta^b = 1 \tag{4}$$

where $b = (1+\sin\phi)/(1-\sin\phi)$. Any orthogonal net satisfying (4) determines a net of principal stress trajectories giving a solution to the Mohr-Coulomb yield criterion and the equilibrium equations. Then, it has been shown in [5] that the flow rule (3) at $\Omega = 0$ and the ideal flow condition are satisfied if

$$uh_\eta = 1 \tag{5}$$

where u is the magnitude of the velocity vector. According to the ideal flow condition the velocity vector is tangent to the $\xi -$ coordinate curves. It is evident that the problem of finding an ideal flow solution is equivalent to the problem of finding a coordinate system satisfying (4). It follows from the geometry of Figure 1 that

$$\frac{\partial x}{\partial \xi} = h_\xi \cos\psi, \quad \frac{\partial x}{\partial \eta} = -h_\eta \sin\psi, \quad \frac{\partial y}{\partial \xi} = h_\xi \sin\psi, \quad \frac{\partial y}{\partial \eta} = h_\eta \cos\psi. \tag{6}$$

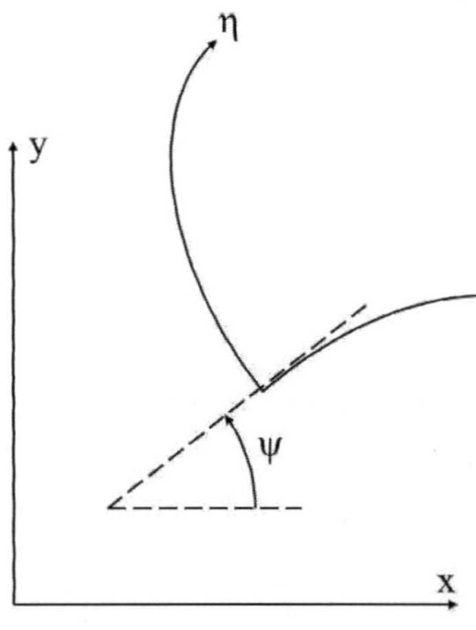

Figure 1. Principal lines and Cartesian coordinates.

The compatibility equations are

$$\frac{\partial^2 x}{\partial \xi \partial \eta} = \frac{\partial^2 x}{\partial \eta \partial \xi}, \quad \frac{\partial^2 y}{\partial \xi \partial \eta} = \frac{\partial^2 y}{\partial \eta \partial \xi}. \tag{7}$$

Substituting (6) into (7) and using (4) and (5) yields

$$bu^{b-1}\cos\psi\frac{\partial u}{\partial\eta}-u^{b}\sin\psi\frac{\partial\psi}{\partial\eta}=\frac{\sin\psi}{u^{2}}\frac{\partial u}{\partial\xi}-\frac{\cos\psi}{u}\frac{\partial\psi}{\partial\xi},$$
$$bu^{b-1}\sin\psi\frac{\partial u}{\partial\eta}+u^{b}\cos\psi\frac{\partial\psi}{\partial\eta}=-\frac{\cos\psi}{u^{2}}\frac{\partial u}{\partial\xi}-\frac{\sin\psi}{u}\frac{\partial\psi}{\partial\xi}. \tag{8}$$

It is always possible to rotate the Cartesian coordinate system so that its x- axis is tangent to the $\xi-$ coordinate curve at a given point. In this case $\psi=0$ and equation (8) becomes

$$bu^{b}\frac{\partial u}{\partial\eta}+\frac{\partial\psi}{\partial\xi}=0,\quad u^{b+2}\frac{\partial\psi}{\partial\eta}+\frac{\partial u}{\partial\xi}=0. \tag{9}$$

Using a standard procedure it is possible to show that this system of equations is hyperbolic with the characteristics

$$\frac{d\eta}{d\xi}=-\sqrt{b}u^{b+1},\quad \frac{d\eta}{d\xi}=\sqrt{b}u^{b+1}. \tag{9}$$

Here the first equation determines the family of $\alpha-$ curves and the second equation the family of $\beta-$ curves. The characteristic relations are

$$\sqrt{b}du-ud\psi=0,\quad \sqrt{b}du+ud\psi=0. \tag{10}$$

Here the first equation is valid on the $\alpha-$ curves and the second equation on the $\beta-$ curves. It follows from (4), (5), and (9) that the angle between the $\xi-$ direction and each of the characteristic directions is $\pi/4+\phi/2$ (Figure 2). Equations (10) can be immediately integrated to give

$$\sqrt{b}\ln u-\psi=-g_{1}(\beta),\quad \sqrt{b}\ln u+\psi=-g_{2}(\alpha). \tag{11}$$

Here $g_2(\alpha)$ is an arbitrary function of α and $g_1(\beta)$ is an arbitrary function of β. Solving (11) for u and ψ yields

$$2\sqrt{b}\ln u=g_{2}(\alpha)-g_{1}(\beta),\quad 2\psi=g_{2}(\alpha)+g_{1}(\beta). \tag{12}$$

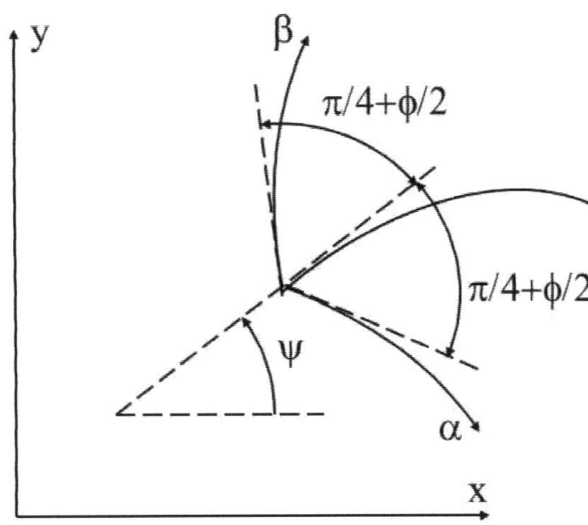

Figure 2. Characteristic coordinates.

In what follows, it is assumed that the characteristic lines of one family are straight. This is an important class of solutions. In particular, the rigid plastic boundary is a straight line if the motion of the rigid zone is a translation. For definitiveness, assume that the $\alpha-$ lines are straight. Let φ be the angle between the $x-$axis and the $\alpha-$ direction measured from the $x-$axis anti-clockwise. Then, it is evident that $\psi=\varphi+\pi/4+\phi/2$ and that the angle φ is independent of α. Therefore, the angle ψ is also independent of α and it follows from (12) that $g_2(\alpha)=2g_0$ is constant. Different choices of the function $g_1(\beta)$ merely change the scale of the $\beta-$curves. Therefore, without loss of generality it is possible to choose that $g_1(\beta)=2\beta$. Hence equation (12) becomes

$$\sqrt{b}\ln u = g_0-\beta, \quad \psi = g_0+\beta. \tag{13}$$

5. Ideal flow solution for tool design

A schematic diagram of the process is shown in Figure 3. There are two rigid regions and one plastic region. The motion of each rigid region is a translation. Therefore, the rigid plastic boundaries, *OC* and *AB*, are straight lines. The channel is converging in the direction of flow. Therefore, it follows from the geometry of Figures 2 and 3 that the rigid plastic boundaries are $\alpha-$ lines and equation (13) is valid in the plastic region. The tool surface *OA* and the thickness of the sheet at the exit are prescribed and the tool surface *BC* that produces an ideal flow should be found from the solution. The origin of the Cartesian coordinate system is chosen at point *O*. The solution can be found by a purely geometric method. The equation of curve *OA* is given in parametric form as

$$x=X(\xi), \quad y=Y(\xi). \tag{14}$$

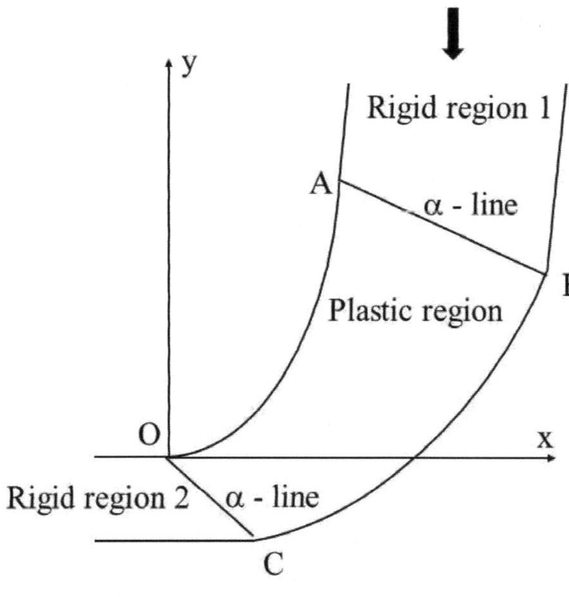

Figure 3. Schematic diagram of the process.

Without loss of generality, it is possible to assume that $\xi=0$ at point *O*. Then,

$$X=0 \quad \text{and} \quad Y=0 \tag{15}$$

at $\xi=0$. Since curve *OA* is an $\xi-$ line, the orientation of the straight $\alpha-$ lines in the plastic region is known (Figure 2). In particular, the equation of these lines is

$$y-Y(\xi)=\left[x-X(\xi)\right]\tan\left(\psi_{OA}-\frac{\pi}{4}-\frac{\phi}{2}\right) \tag{16}$$

where ψ_{OA} is the value of ψ on curve OA. Using trigonometric relations equation (16) can be transformed to

$$y - Y(\xi) = \left[x - X(\xi)\right]\left(\frac{\tan\psi_{OA} - m}{1 + m\tan\psi_{OA}}\right) \tag{17}$$

where $m = \tan(\pi/4 + \phi/2)$. Using (14) it is possible to find that

$$\tan\psi_{OA} = \left(\frac{dY}{d\xi}\right)\left(\frac{dX}{d\xi}\right)^{-1} = \lambda(\xi). \tag{18}$$

The curve BC is also an $\xi-$ line. Therefore, the equation of this curve is

$$\frac{dy}{dx} = \tan\psi_{OA} = \lambda(\xi). \tag{19}$$

Differentiating (17) with respect to ξ and using (18) leads to

$$\frac{dy}{d\xi} - \frac{dY}{d\xi} = \left(\frac{dx}{d\xi} - \frac{dX}{d\xi}\right)\left(\frac{\lambda - m}{1 + m\lambda}\right) + (x - X)\Lambda \tag{20}$$

where

$$\Lambda(\xi) = \frac{d\lambda}{d\xi}\frac{(1 + m^2)}{(1 + m\lambda)^2}. \tag{21}$$

Equation (19) can be rewritten as

$$\frac{dy}{d\xi} = \lambda\frac{dx}{d\xi}. \tag{22}$$

Eliminating the derivative $dy/d\xi$ in (20) by means of (22) yields

$$m\frac{dx}{d\xi} - x\Lambda\left(\frac{1 + m\lambda}{\lambda^2 + 1}\right) = \left(\frac{dY}{d\xi} - X\Lambda\right)\left(\frac{1 + m\lambda}{\lambda^2 + 1}\right) - \left(\frac{\lambda - m}{\lambda^2 + 1}\right)\frac{dX}{d\xi}. \tag{23}$$

This is a linear differential equation for x. Its general solution can be found without any difficulty. Using (19) equation (17) can be rewritten as

$$x - X = (y - Y)\left(\frac{1 + m\lambda}{\lambda - m}\right) \tag{24}$$

Differentiating this equation with respect to ξ leads to

$$\frac{dx}{d\xi} - \frac{dX}{d\xi} = \left(\frac{dy}{d\xi} - \frac{dY}{d\xi}\right)\left(\frac{1 + m\lambda}{\lambda - m}\right) + (y - Y)P \tag{25}$$

where

$$P(\xi) = -\frac{d\lambda}{d\xi}\frac{(1 + m^2)}{(\lambda - m)^2}. \tag{26}$$

Eliminating the derivative $dx/d\xi$ in (25) by means of (22) yields

IOP Conf. Series: Materials Science and Engineering **311** (2018) 012013 doi:10.1088/1757-899X/311/1/012013

$$m\frac{dy}{d\xi}-yP\lambda\left(\frac{m-\lambda}{\lambda^2+1}\right)=\lambda\left(\frac{dX}{d\xi}-YP\right)\left(\frac{m-\lambda}{\lambda^2+1}\right)+\lambda\left(\frac{1+m\lambda}{\lambda^2+1}\right)\frac{dY}{d\xi}. \tag{27}$$

This is a linear differential equation for y. The solution of equations (23) and (27) determines the shape of BC. The thickness of the sheet at the exit is denoted by H. Therefore, the boundary conditions to equations (23) and (27) are

$$x=\frac{H\sin(\pi/4-\phi/2+\psi_0)}{\cos(\pi/4-\phi/2)} \quad \text{and} \quad y=-\frac{H\cos(\pi/4-\phi/2+\psi_0)}{\cos(\pi/4-\phi/2)} \tag{28}$$

at $\xi=0$, respectively. Here ψ_0 is the value of ψ_{OA} at $\xi=0$.

6. Illustrative example

Assume that curve OA is a parabola (Figure 3). Taking into account (14) and (15) its equation can be written as

$$X(\xi)=\xi, \quad Y(\xi)=n\xi^2 \tag{29}$$

where n is constant. Then, it follows from (18), (21) and (26) that

$$\lambda=2n\xi, \quad \Lambda=\frac{2n(1+m^2)}{(1+2mn\xi)^2}, \quad P=-\frac{2n(1+m^2)}{(2n\xi-m)^2}. \tag{30}$$

Substituting (29) and (30) into (23) yields

$$\frac{dx}{d\xi}-\frac{2n(1+m^2)x}{m(4n^2\xi^2+1)(1+2nm\xi)}=1+\frac{1}{(1+2mn\xi)}-\frac{(m+2n\xi)}{m(1+4n^2\xi^2)}. \tag{31}$$

It follows from (30) that $\psi_0=0$. The solution of equation (31) satisfying the boundary condition (28) at $\psi_0=0$ is

$$x=\frac{(1+2mn\xi)}{\sqrt{1+4n^2\xi^2}}\exp\left[\frac{\arctan(2n\xi)}{m}\right]\times$$
$$\left\{H\tan\left(\frac{\pi}{4}-\frac{\phi}{2}\right)+\int_0^\xi\frac{\sqrt{1+4n^2z^2}}{(1+2mnz)}\exp\left[-\frac{\arctan(2nz)}{m}\right]\left[1+\frac{1}{(1+2mnz)}-\frac{(m+2nz)}{m(1+4n^2z^2)}\right]dz\right\}. \tag{32}$$

Here z is a dummy variable of integration. Substituting (29) and (30) into (27) yields

$$\frac{dy}{d\xi}+\frac{4n^2(1+m^2)\xi y}{m(4n^2\xi^2+1)(m-2n\xi)}=\frac{2n\xi\left[m^2-2nm\xi+2(1+3m^2)n^2\xi^2-8mn^3\xi^3\right]}{m(1+4n^2\xi^2)(m-2n\xi)}. \tag{33}$$

The solution of this equation satisfying the boundary condition (28) at $\psi_0=0$ is

$$y=\frac{2n}{m}\frac{\sqrt{1+4n^2\xi^2}}{(m-2n\xi)}\exp\left[-\frac{\arctan(2n\xi)}{m}\right]\times$$
$$\left\{\int_0^\xi z\exp\left[\frac{\arctan(2nz)}{m}\right]\frac{\left[m^2-2nmz+2(1+3m^2)n^2z^2-8mn^3z^3\right]}{(1+4n^2z^2)^{3/2}}dz-\frac{Hm}{2n}\right\}. \tag{34}$$

Equations (32) and (34) determine curve BC in parametric form with ξ being the parameter.

7. Conclusions

The theory of planar bulk ideal flows for tool design has been developed assuming that one family of characteristic curves is straight. The theory is valid for both the double – shearing model [16] and the double slip and rotation model [7] of pressure-dependent plasticity. Finding an optimal tool shape has been reduced to solving linear ordinary differential equations. An illustrative example has been given assuming that one surface of tool is determined by equation (29) and that the motion of each rigid region is a translation. Then, the other surface is determined from the solution and is given by equations (32) and (34). It is seen from these equations that the solution is practically analytic. A numerical technique is only necessary to evaluate ordinary integrals.

8. References

[1] Chung K and Richmond O 1994 *ASME J. Appl. Mech.* **61** 176
[2] Richmond O and Alexandrov S 2002 *Acta Mech.* **158** 33
[3] Chung K and Alexandrov S 2007*Appl. Mech. Rev.* **60** 316
[4] Alexandrov S, Mustafa Y and Lyamina E 2016 *Meccanica* **51** 2235
[5] Alexandrov S 2017 *Poromechanics VI: Proc. 6th Biot Conf. on Poromechanics (Reston: ASCE)* p 967
[6] Collins I F and Meguid S A 1977 *ASME J. Appl. Mech.* **44** 271
[7] Harris D 2014 *Acta Mech.* **225** 2277
[8] Richmond O and Devenpeck M L 1962 *Proc. 4th U.S. Nat. Congr. on Appl. Mech. (New York: ASME)* p 1053
[9] Richmond O and Morrison H L 1967 *J. Mech. Phys. Solids* **15** 195
[10] Richmond O 1968 *Mechanics of the Solid State*, ed F P J Rimrott and J Schwaighofer *(Toronto: University of Toronto Press)* p 154
[11] Alexandrov S, Lee W, Chung K and Kang T J 2004 *Acta Mech.* **173** 49
[12] Chung K H, Chung K and Alexandrov S 2008 *Acta Mech.* **196** 127
[13] Alexandrov S and Date P 2017 *MATEC Web Conf. 2017*, vol 108 Article 01001
[14] Alexandrov S 2003 *Acta Mech.* **163** 127
[15] Lyamina E and Alexandrov S 2016 *Key Eng. Mater.* **713** 143
[16] Spencer A J M 1964 *J. Mech. Phys. Solids* **12** 337
[17] Spitzig W A, Sober R J and Richmond O 1976 *Metallurg. Trans. A* **7** 1703
[18] Richmond O and Alexandrov S 2000 *J. Mech. Phys. Solids* **48** 1735
[19] Alexandrov S 1999 *Proc. Conf. on the 70th Birthday of Dr. Owen Richmond "The Integration of Material, Process and Product Design" (Rotterdam: Balkema)* p 103
[20] Hill R 1967 *J. Mech. Phys. Solids* **15** 223
[21] Alexandrov S and Harris D 2017 *Z. Angew. Math. Mech.* **97** 473

Singular solutions for the rigid plastic double slip and rotation model under plane strain

S Alexandrov[1,2] and E Lyamina[3]

1 Visiting Professor, Division of Computational Mathematics and Engineering, Institute for Computational Science, Ton Duc Thang University, Ho Chi Minh City, Vietnam
2 Visiting Professor, Faculty of Civil Engineering, Ton Duc Thang University, Ho Chi Minh City, Vietnam
3 Senior Researcher, Laboratory of Strenght and Fracture of Materials and Structures, Institute for Problems in Mechanics, Moscow, Russia
E-mail: sergeialexandrov@tdt.edu.vn

Abstract. In the mechanics of granular and other materials the system of equations comprising the rigid plastic double slip and rotation model together with the stress equilibrium equations under plane strain conditions forms a hyperbolic system. Boundary value problems for this system of equations can involve a frictional interface. An envelope of characteristics may coincide with this interface. In this case, the solution is singular. In particular, some components of the strain rate tensor approach infinity in the vicinity of the frictional interface. Such behavior of solutions is in qualitative agreement with experimental data that show that a narrow layer of localized plastic deformation is often generated near frictional interfaces. The present paper deals with asymptotic analysis of the aforementioned system of equations in the vicinity of an envelope of characteristics. It is shown that the shear strain rate and the spin component in a local coordinate system connected to the envelope follow an inverse square root rule in its vicinity.

1. Introduction

Singular solutions have been studied for several rigid plastic models. In particular, the rigid perfectly plastic model has been considered [1]. In this work, three dimensional flow of material obeying an arbitrary smooth yield criterion and its associated flow rule has been analysed. It has been shown that the singular solutions appear near surfaces on which the shear stress is equal to the shear yield stress and that the quadratic invariant of the strain rate tensor follows an inverse square root rule near the surface. This result has been extended to a class of viscoplastic models in [2] for planar flow and in [3] for axisymmetric flow. The constitutive equations of this class of models include a saturation stress. In particular, the yield stress in tension approaches the saturation stress as the quadratic invariant of the strain rate tensor approaches infinity. The qualitative asymptotic behaviour of solutions is controlled by the exact asymptotic behaviour of the dependence of the yield stress on the quadratic invariant of the strain rate tensor at infinity. For a certain class of functions the quadratic invariant of the strain rate tensor follows an inverse square root rule near the surface where the shear stress is equal to the shear saturation stress. The model of anisotropic plasticity proposed in [4] has been investigated in [5] under plane strain conditions. The corresponding system of equations is hyperbolic. The singularity in solutions develops near envelopes of characteristics. The exact asymptotic singular behaviour of solutions depends on the shape of the yield surface near the point corresponding to the state of stress at the envelope. For a certain class of functions the quadratic invariant of the strain rate tensor follows an inverse square root rule near the envelope. The present paper deals with plane strain deformation of materials that obeys the rigid plastic double slip and rotation model proposed in [6]. The corresponding system of equations is hyperbolic. It is shown that the shear strain rate and the spin component in a local coordinate system connected to an envelope of characteristics follow an inverse square root rule in its vicinity.

The technological value of the present research is that a narrow layer of localized plastic deformation is generated near frictional interfaces in machining and deformation processes ([7]-[16] among many others). These layers affect the performance of machine parts [17]. The main theoretical result reported in [1]-[3], [5] as well as in the present paper for the rigid plastic double slip and rotation model demonstrates that the models considered in these papers are capable of predicting the generation of the layer of localized plastic deformation near frictional interfaces.

2. The rigid plastic double slip and rotation model under plain strain

The double slip and rotation model has been proposed in [6]. Under plane strain conditions of incompressible material, the rigid plastic constitutive equations of the model are the Mohr - Coulomb yield criterion (other yield criteria may be adopted as well) and the flow rule. In an arbitrary Cartesian coordinate system (x, y) these equations can be written as

$$\left(\sigma_{xx}+\sigma_{yy}\right)\sin\phi+\sqrt{\left(\sigma_{xx}-\sigma_{yy}\right)^2+4\sigma_{xy}^2}=2k\cos\phi \tag{1}$$

and

$$\varepsilon_{xx}+\varepsilon_{yy}=0,\quad \sin 2\psi\left(\varepsilon_{xx}-\varepsilon_{yy}\right)-2\cos 2\psi\varepsilon_{xy}-2\sin\phi\left(\omega_{xy}+\Omega\right)=0. \tag{2}$$

Here σ_{xx}, σ_{yy} and σ_{xy} are the components of the stress tensor referred to the (x, y) coordinate system, ε_{xx}, ε_{yy} and ε_{xy} are the components of the strain rate tensor referred to the (x, y) coordinate system, ω_{xy} is the only non-zero spin (vorticity) component referred to the (x, y) coordinate system, ψ is the angle between the $x-$ direction and the greatest principal stress σ_1 measured from the $x-$ direction anti-clockwise, Ω is the intrinsic spin due to grain rotation, k is the cohesion, and ϕ is the angle of internal friction. The quantity Ω is an unknown variable which is governed by the equation of rotational motion. It is often reasonable to suppose that in the interior of the yielding region the grain rotation is either zero or averages to zero [18]. In the present paper, it is assumed that Ω is constant.

3. Characteristics, characteristic relations and envelopes

The system of equations comprising (1), (2) and the equations of equilibrium is hyperbolic. Let u_x and u_y be the velocity components referred to the Cartesian coordinate system. Then, equation (2) can be rewritten as

$$\begin{gathered}\frac{\partial u_x}{\partial x}+\frac{\partial u_y}{\partial y}=0,\\ \sin 2\psi\left(\frac{\partial u_x}{\partial x}-\frac{\partial u_y}{\partial y}\right)-\left(\cos 2\psi+\sin\phi\right)\frac{\partial u_x}{\partial y}-\left(\cos 2\psi-\sin\phi\right)\frac{\partial u_y}{\partial x}-2\Omega\sin\phi=0.\end{gathered} \tag{3}$$

The components of the stress tensor can be represented as [19]

$$\sigma_{xx}=-p+q\cos 2\psi,\quad \sigma_{yy}=-p-q\cos 2\psi,\quad \sigma_{xy}=q\sin 2\psi \tag{4}$$

where p and q are the stress invariant defined as

$$p = -\frac{\left(\sigma_{xx} + \sigma_{yy}\right)}{2}, \quad q = \sqrt{\frac{1}{4}\left(\sigma_{xx} - \sigma_{yy}\right)^2 + \sigma_{xy}^2}. \tag{5}$$

Then, the yield criterion (1) becomes

$$-p\sin\phi + q = k\cos\phi. \tag{6}$$

The equilibrium equations are

$$\frac{\partial \sigma_{xx}}{\partial x} + \frac{\partial \sigma_{xy}}{\partial y} = 0, \quad \frac{\partial \sigma_{xy}}{\partial x} + \frac{\partial \sigma_{yy}}{\partial y} = 0. \tag{7}$$

Substituting (4) into (7) yields

$$\begin{aligned} -\frac{\partial p}{\partial x} + \cos 2\psi \frac{\partial q}{\partial x} - 2q\sin 2\psi \frac{\partial \psi}{\partial x} + \sin 2\psi \frac{\partial q}{\partial y} + 2q\cos 2\psi \frac{\partial \psi}{\partial y} = 0, \\ \sin 2\psi \frac{\partial q}{\partial x} + 2q\cos 2\psi \frac{\partial \psi}{\partial x} - \frac{\partial p}{\partial y} - \cos 2\psi \frac{\partial q}{\partial y} + 2q\sin 2\psi \frac{\partial \psi}{\partial y} = 0. \end{aligned} \tag{8}$$

In these equations, p can be eliminated by means on (6). The resulting system contains two unknowns, q and ψ. This system is hyperbolic and the angle between the direction of the greatest principal stress σ_1 and the characteristic directions is $\pm(\pi/4 + \phi/2)$ [19]. The Cartesian (x, y) and characteristic (ξ, η) coordinate systems are illustrated in Fig. 1. It is always possible to rotate the Cartesian coordinate system such that its $x-$axis is tangent to an $\xi-$ characteristic curve at a given point. Then, $\psi = \pi/4 + \phi/2$ at this point and equation (3) becomes

$$\frac{\partial u_x}{\partial x} + \frac{\partial u_y}{\partial y} = 0, \quad \cos\phi \frac{\partial u_x}{\partial x} + \sin\phi \frac{\partial u_y}{\partial x} - \Omega \sin\phi = 0. \tag{9}$$

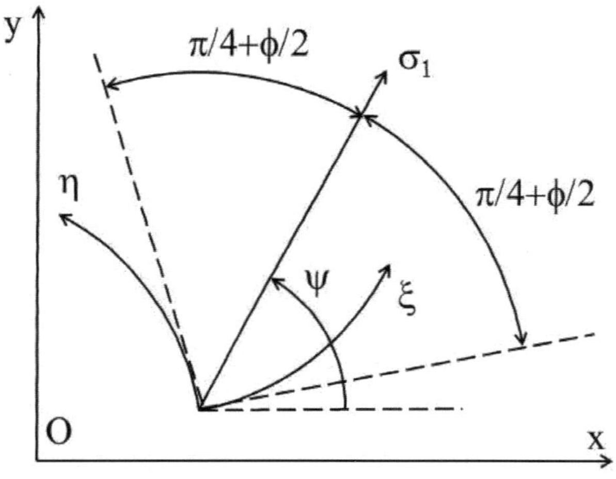

Figure 1. Cartesian and characteristic coordinates.

In the second equation of this system the derivative $\partial u_y/\partial y$ has been eliminated by means of the first equation. The second equation in (9) is one of the characteristic relations. This equation is valid if $|\partial u_x/\partial y| < \infty$. In order to study the behaviour of solutions in the vicinity of an envelope of $\xi-$ characteristic curves, it is necessary to assume that

IOP Conf. Series: Materials Science and Engineering **311** (2018) 012014 doi:10.1088/1757-899X/311/1/012014

$$\left|\frac{\partial u_x}{\partial y}\right| \to \infty \tag{10}$$

as $y \to 0$. In this case the second equation in (9) is not valid because

$$(\cos 2\psi + \sin\phi)\frac{\partial u_x}{\partial y} \neq 0 \tag{11}$$

at $y = 0$. Eliminating the derivative $\partial u_y/\partial y$ in the second equation in (3) by means of the first equation yields

$$2\sin 2\psi \frac{\partial u_x}{\partial x} - (\cos 2\psi + \sin\phi)\frac{\partial u_x}{\partial y} - (\cos 2\psi - \sin\phi)\frac{\partial u_y}{\partial x} - 2\Omega\sin\phi = 0. \tag{12}$$

In what follows, it is assumed that the in-surface derivatives are bounded. Then, it follows from (12) that

$$\left|(\cos 2\psi + \sin\phi)\frac{\partial u_x}{\partial y}\right| < \infty \tag{13}$$

at $y = 0$.

4. Asymptotic analysis

Consider the solution behaviour of equations (3) and (8) in the vicinity of the maximum friction surface assuming that equation (11) is satisfied. It is convenient to adopt the Cartesian coordinate system whose *x*- axis is tangent to the maximum friction surface at a given point (i.e. $\psi = \pi/4 + \phi/2$ at this point) and whose origin is situated at this point (i.e. $y = 0$ at this point). In what follows, it is assumed that the solution is represented by power series of *y* near $y = 0$. Then, the velocity component u_x is represented as

$$u_x = u_0 + u_1 y^\alpha + u_2 y + o(y) \tag{14}$$

as $y \to 0$. Here u_0 and u_1 are independent of *y*. Since the velocity component u_x must be bounded, it is obvious from (12) that $\alpha > 0$. On the other hand, equation (11) demands $\alpha < 1$. Therefore,

$$0 < \alpha < 1. \tag{15}$$

The angle ψ is represented as

$$\psi = \frac{\pi}{4} + \frac{\phi}{2} + \psi_0 y^\beta + \psi_1 y + o(y) \tag{16}$$

as $y \to 0$. Here ψ_0 and ψ_1 are independent of *y* and $\beta > 0$. Substituting (14) and (16) into (11) and taking into account (13) gives $\alpha + \beta - 1 = 0$ or

$$\beta = 1 - \alpha. \tag{17}$$

It is seen from (15) and (17) that $0 < \beta < 1$. Then, it follows from (16) and (17) that

$$\begin{aligned}
&\cos 2\psi = -\sin\phi - 2\cos\phi\left(\psi_0 y^{1-\alpha} + \psi_1 y\right) + o(y),\\
&\sin 2\psi = \cos\phi - 2\sin\phi\left(\psi_0 y^{1-\alpha} + \psi_1 y\right) + o(y),\\
&\cos 2\psi + \sin\phi = -2\cos\phi\left(\psi_0 y^{1-\alpha} + \psi_1 y\right) + o(y),\\
&\cos 2\psi - \sin\phi = -2\sin\phi - 2\cos\phi\left(\psi_0 y^{1-\alpha} + \psi_1 y\right) + o(y)
\end{aligned} \tag{18}$$

as $y \to 0$. Using (14) the first equation in (3) can be represented as

$$\frac{\partial u_y}{\partial y} = -\frac{du_0}{dx} + \frac{du_1}{dx} y^\alpha + \frac{du_1}{dx} y + o(y) \tag{19}$$

as $y \to 0$. Integrating this equation with respect to y leads to

$$u_y = -\frac{du_0}{dx} y + o(y) \tag{20}$$

as $y \to 0$. It has been taken into account here that it is always possible to assume with no loss of generality that the rigid tool is motionless and, therefore, $u_y = 0$ at $y = 0$. Using (14), (18) and (20) the terms involved in (12) are represented as

$$\begin{aligned}
&2\sin 2\psi \frac{\partial u_x}{\partial x} = 2\cos\phi \frac{du_0}{dx} + 2\cos\phi \frac{du_1}{dx} y^\alpha - 4\psi_0 \sin\phi \frac{du_0}{dx} y^{1-\alpha} + O(y),\\
&\left(\cos 2\psi + \sin\phi\right)\frac{\partial u_x}{\partial y} = -2\psi_0 \alpha u_1 \cos\phi - 2\psi_0 u_2 \cos\phi\, y^{1-\alpha} - 2\psi_1 \alpha u_1 \cos\phi\, y^\alpha + O(y),\\
&\left(\cos 2\psi - \sin\phi\right)\frac{\partial u_y}{\partial x} = O(y)
\end{aligned} \tag{21}$$

as $y \to 0$. Substituting (21) into (12) gives

$$\begin{aligned}
&2\left(\cos\phi \frac{du_0}{dx} + \psi_0 \alpha u_1 \cos\phi - \Omega \sin\phi\right) + 2\cos\phi\left(\frac{du_1}{dx} + \psi_1 \alpha u_1\right) y^\alpha +\\
&2\psi_0\left(u_2 \cos\phi - 2\sin\phi\right) y^{1-\alpha} + O(y)
\end{aligned} \tag{22}$$

as $y \to 0$. It is seen from this equation that $1 - \alpha = \alpha$ or

$$\alpha = \frac{1}{2}. \tag{23}$$

It remains to demonstrate that this result does not contradict the stress equations. Eliminating p in (8) by means of (6) yields

$$\begin{aligned}
&\left(\cos 2\psi - \frac{1}{\sin\phi}\right)\frac{\partial q}{\partial x} - 2q\sin 2\psi \frac{\partial \psi}{\partial x} + \sin 2\psi \frac{\partial q}{\partial y} + 2q\cos 2\psi \frac{\partial \psi}{\partial y} = 0,\\
&\sin 2\psi \frac{\partial q}{\partial x} + 2q\cos 2\psi \frac{\partial \psi}{\partial x} - \left(\frac{1}{\sin\phi} + \cos 2\psi\right)\frac{\partial q}{\partial y} + 2q\sin 2\psi \frac{\partial \psi}{\partial y} = 0.
\end{aligned} \tag{24}$$

Using (17) and (23) it is possible to rewrite equations (16) and (18) in the form

$$\begin{aligned}
&\psi=\frac{\pi}{4}+\frac{\phi}{2}+\psi_0\sqrt{y}+\psi_1 y+o(y),\\
&\cos 2\psi=-\sin\phi-2\cos\phi\left(\psi_0\sqrt{y}+\psi_1 y\right)+o(y),\\
&\sin 2\psi=\cos\phi-2\sin\phi\left(\psi_0\sqrt{y}+\psi_1 y\right)+o(y)
\end{aligned}\tag{25}$$

as $y\to 0$. Then,

$$\begin{aligned}
&\cos 2\psi-\frac{1}{\sin\phi}=-\frac{1+\sin^2\phi}{\sin\phi}-2\cos\phi\left(\psi_0\sqrt{y}+\psi_1 y\right)+o(y),\\
&\cos 2\psi+\frac{1}{\sin\phi}=\frac{\cos^2\phi}{\sin\phi}-2\cos\phi\left(\psi_0\sqrt{y}+\psi_1 y\right)+o(y)
\end{aligned}\tag{26}$$

as $y\to 0$. It is seen from (25) that

$$q\cos 2\psi\frac{\partial\psi}{\partial y}=O\left(\frac{1}{\sqrt{y}}\right),\quad q\sin 2\psi\frac{\partial\psi}{\partial y}=O\left(\frac{1}{\sqrt{y}}\right)\tag{27}$$

as $y\to 0$. Therefore, it follows from (24), (25) and (26) that $\partial q/\partial y=O\left(1/\sqrt{y}\right)$ as $y\to 0$. Then, the function q can be represented as

$$q=q_0+q_1\sqrt{y}+q_2 y+o(y)\tag{28}$$

as $y\to 0$. Here q_0, q_1 and q_2 are independent of y. It follows from (25), (26) and (28) that

$$\begin{aligned}
&\left(\cos 2\psi-\frac{1}{\sin\phi}\right)\frac{\partial q}{\partial x}=-\frac{\left(1+\sin^2\phi\right)}{\sin\phi}\frac{dq_0}{dx}-\left[\frac{\left(1+\sin^2\phi\right)}{\sin\phi}\frac{dq_1}{dx}+2\psi_0\cos\phi\frac{dq_0}{dx}\right]\sqrt{y}+O(y),\\
&-2q\sin 2\psi\frac{\partial\psi}{\partial x}=q\frac{\partial\cos 2\psi}{\partial x}=-2q_0\cos\phi\frac{d\psi_0}{dx}\sqrt{y}+O(y),\\
&\sin 2\psi\frac{\partial q}{\partial y}=\frac{q_1\cos\phi}{2\sqrt{y}}+\left(q_2\cos\phi-q_1\psi_0\sin\phi\right)+O\left(\sqrt{y}\right),\\
&2q\cos 2\psi\frac{\partial\psi}{\partial y}=q\frac{\partial\sin 2\psi}{\partial y}=-\frac{q_0\psi_0\sin\phi}{\sqrt{y}}+\left(2\psi_1 q_0+\psi_0 q_1\right)\sin\phi+O\left(\sqrt{y}\right),\\
&\sin 2\psi\frac{\partial q}{\partial x}=\cos\phi\frac{dq_0}{dx}+\left(\cos\phi\frac{dq_1}{dx}-2\psi_0\sin\phi\frac{dq_0}{dx}\right)\sqrt{y}+O(y),\\
&2q\cos 2\psi\frac{\partial\psi}{\partial x}=q\frac{\partial\sin 2\psi}{\partial x}=-2q_0\sin\phi\frac{d\psi_0}{dx}\sqrt{y}+O(y),\\
&-\left(\frac{1}{\sin\phi}+\cos 2\psi\right)\frac{\partial q}{\partial y}=-\frac{q_1\cos^2\phi}{2\sin\phi\sqrt{y}}+\left(q_1\psi_0\cos\phi-\frac{q_2\cos^2\phi}{\sin\phi}\right)+O\left(\sqrt{y}\right),\\
&2q\sin 2\psi\frac{\partial\psi}{\partial y}=-q\frac{\partial\cos 2\psi}{\partial y}=\frac{q_0\psi_0\cos\phi}{\sqrt{y}}+\left(2q_0\psi_1+q_1\psi_0\right)\cos\phi+O\left(\sqrt{y}\right)
\end{aligned}\tag{29}$$

as $y\to 0$. It is seen from (24) and (28) that each of the equations in (24) reduces to the equation of the form $A+B\sqrt{y}+o\left(\sqrt{y}\right)$ as $y\to 0$ where A and B are independent of y if

$$q_1\cot\phi-2q_0\psi_0=0.\tag{30}$$

In this case, equation (24) is compatible with (23) and the asymptotic representation of the velocity component u_x in the vicinity of the envelope (or the maximum friction surface) follows from (14) and (23) in the form

$$u_x = u_0 + u_1\sqrt{y} + o\left(\sqrt{y}\right) \tag{31}$$

as $y \to 0$. It is seen from this equation that $\partial u_x / \partial y = O\left(1/\sqrt{y}\right)$ as $y \to 0$ and, therefore, both $\varepsilon_{xy} = O\left(1/\sqrt{y}\right)$ and $\omega_{xy} = O\left(1/\sqrt{y}\right)$ as $y \to 0$.

5. Conclusions

It has been shown that plane strain rigid plastic solutions for the double slip and rotation model for incompressible material are singular (in the sense that some space derivatives of velocities and stresses approach infinity) in the vicinity of maximum friction surfaces. The maximum friction surface is defined by the condition that an envelope of characteristics coincides with the surface. The exact asymptotic expansion of velocities, strain rates and stresses in the vicinity of singular surfaces has been found. It has been hypothesized that this singular behaviour of the velocity field can be adopted for describing the evolution of material properties in a narrow layer near frictional interfaces cause by highly localized plastic deformation.

References

[1] Alexandrov S and Richmond O 2001 *Int. J. Non-Linear Mech.* **36** 1
[2] Alexandrov S and Mustafa Y 2013 *Meccanica* **48** 2203
[3] Alexandrov S and Mustafa Y 2015 *Appl. Math. Model.* **39** 4599
[4] Rice J R 1973 *J. Mech. Phys. Solids* **21** 63
[5] Alexandrov S and Jeng Y-R 2013 *Cont. Mech. Therm.* **25** 685
[6] Harris D 2014 *Acta Mech.* **225** 2277
[7] Griffiths B J 1987 *Trans. ASME J. Trib.* **109** 525
[8] Kim Y-T and Ikeda K 2000 *Metallurg. Mater. Trans.-A* **31** 1635
[9] Murai T, Matsuoka S, Miyamoto S and Oki Y 2003 *J. Mater. Process. Technol.* **141** 207
[10] Kajino S and Asakawa M 2006 *J. Mater. Process. Technol.* **177** 704
[11] Umbrello D. and Jawahir I S 2009 *Int. J. Adv. Manuf. Technol.* **44** 955
[12] Sasaki T T, Morris R A, Thompson G B, Syarif Y and Fox D 2010 *Scripta Mater.* **63** 488
[13] Veldhuis S C, Dosbaeva G K, Elfizy A, Fox-Rabinovich G S and Wagg T 2010 *J. Mater. Eng. Perform.* **19** 1031
[14] Thirumurugan M, Rao S A, Kumaran S and Rao T S 2011 *J. Mater. Process. Technol.* **211** 1637
[15] Li X, Zu G, Ding M, Mu Y and Wang P 2011 *Mater. Sci. Technol-A* **529** 485
[16] Sanabria V and Mueller S 2017 *AIP Conf. Proc.* **1896** 14012
[17] Cho D-H, Lee S-A and Lee Y-Z 2012 *Tribol. Lett.* **45** 123
[18] Alexandrov S and Harris D 2010 *Eur. J. Mech.-A/Solids* **29** 966
[19] Spencer A J M 1964 *J. Mech. Phys. Solids* **12** 337

IOP Conf. Series: Materials Science and Engineering **311** (2018) 012015 doi:10.1088/1757-899X/311/1/012015

One-step microwave plasma enhanced chemical vapor deposition (MW-PECVD) for transparent superhydrophobic surface

Sukrit Thongrom[1,2], Yutthana Tirawanichakul[1], Nantakan Munsit[1,2] and Chalongrat Deangngam[1,2]

1 Department of Physics, Faculty of Science, Prince of Songkla University, Songkhla 90110, Thailand
2 Center of Excellence in Nanotechnology for Energy (CENE), Prince of Songkla University, Songkhla 90110, Thailand
E-mail: chalongrat.d@psu.ac.th

Abstract. We demonstrate a rapid and environmental friendly fabrication technique to produce optically clear superhydrophobic surfaces using poly (dimethylsiloxane) (PDMS) as a sole coating material. The inert PDMS chain is transformed into a 3-D irregular solid network through microwave plasma enhanced chemical vapor deposition (MW-PECVD) process. Thanks to high electron density in the microwave-activated plasma, coating can be done in just a single step with rapid deposition rate, typically much shorter than 10 s. Deposited layers show excellent superhydrophobic properties with water contact angles of ~170° and roll-off angles as small as ~3°. The plasma-deposited films can be ultrathin with thicknesses under 400 nm, greatly diminishing the optical loss. Moreover, with appropriate coating conditions, the coating layer can even enhance the transmission over the entire visible spectrum due to a partial anti-reflection effect.

1. Introduction

Fabrication of transparent superhydrophobic coating is attracting more and more attention owing to numbers of potential applications, such as high-performance solar cells, self-cleaning residential or vehicle glasses, as well as anti-contamination packaging. Essentially, superhydrophobic properties rely strictly on at least two key components: hydrophobic chemicals and a rough surface topology in micro-/nanoscale. Furthermore, it has been proved that effective and stable Cassie-Baxter non-wetting state requires multiscale roughness [1, 2]. However, increased irregularities generally lead to poor optical properties due to substantial light scattering [3]. Therefore, incorporation of transparent property into superhydrophobic coating requires optimized surface roughness in combination with ultrathin layers to minimize light scattering and bulk absorption, respectively [4, 5].

Several industrially viable techniques, for instance, spray coating [6], dip coating [7], or spin coating [8] of hydrophobic nanomaterials on substrates have been demonstrated for the fabrication of ultrathin wetting-resistant layers with low optical loss. Nevertheless, the aforementioned methods usually rely on multi-step and time-consuming processes necessary for nanoparticle synthesis or functionalization. Furthermore, they share common disadvantages on involving many reactive chemicals, especially volatile organic compound (VOC) that may pose health risks. Recently, chemical vapor deposition (CVD) assisted by radio frequency (RF) plasma has been employed to fabricate superhydrophobic coating via plasma polymerization [9]. The CVD process is even much faster when operated under

MW plasma [10]. Although those plasma-based depositions are free from organic solvents the process still employs hazardous and flammable precursors.

In an effort to develop an environmental friendly and fast method for generating transparent superhydrophobic layers, we investigated the feasibility of utilizing MW-PECVD with inert poly (dimethylsiloxane) (PDMS) liquid as the only input material used in the coating step. This organosiloxane compound is well known for its low hazard level, UV-resistance, and anti-corrosion abilities, ideal for use as a coating material [11]. Taking advantage of high electron density and collision frequency provided by MW-activated plasma [12], the PDMS molecules are expected to undergo substantial fragmentation into shorter reactive chains, which eventually polymerize again, thus forming a disordered 3-D structure on the surface. With this suitable surface roughness in combination with a minimum film thickness, superhydrophobicity and transparency can co-exist. Furthermore, the coating itself is a one-step process with a very fast deposition time of less than 10 s.

2. Experiment

2.1 Materials

Poly (dimethylsiloxane) or silicone oil, with a viscosity of 20 cSt was purchased from Sigma-Aldrich. This organosiloxane compound was chosen as a coating material since it contains inorganic backbone and hydrophobic methyl groups. Microscope glass slides (25.4 x 76.2 mm) were obtained from Sail Brand. Pure oxygen was supplied from Linde.

2.2. Plasma reaction system

A setup diagram of a custom-built system capable of performing plasma cleaning and ultrathin film coating via MW-PECVD technique is shown in Figure 1. A commercial microwave oven (800 W, 2.45 GHz) was employed as the plasma excitation source. A round quartz chamber (16 cm diameter and 5 cm height) with a gas inlet and outlet port was placed inside the microwave oven. Vacuum was generated by connecting the gas outlet to a mechanical pump (E2M8, Edwards) via a silicone tube passing through the microwave oven sidewall. Similarly, oxygen gas or vapor of coating precursors was delivered to the reaction chamber via another silicone tube connected to the inlet port. Oxygen flow rate was regulated using a flow meter, whereas the precursor rate is controlled through a heating temperature ranging from 50 °C to 130 °C. Plasma processing temperature in the chamber is typically 150-200 °C.

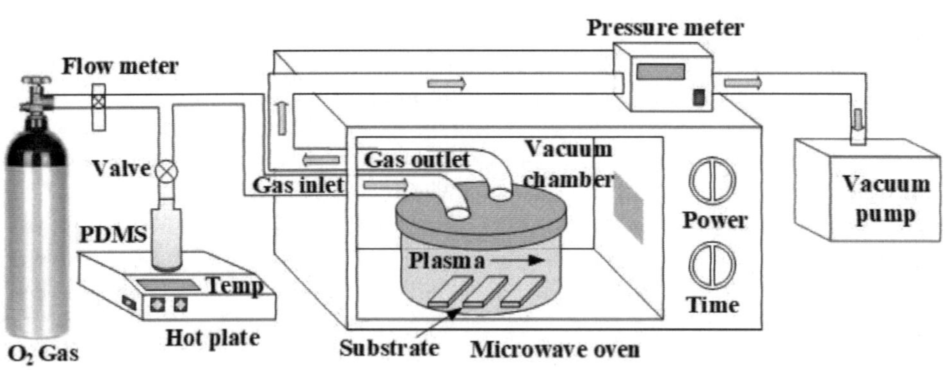

Figure 1. Schematic diagram of a custom MW-PECDV system used in this study.

2.3 Film deposition process

Microscope slides were used as planar substrates and were cleaned with oxygen plasma to remove surface contaminants at molecular level before the film deposition process. Briefly, glass samples were placed inside the chamber, which was then evacuated to a vacuum of ~0.1 mbar. Then oxygen gas was fed into until the chamber pressure reached ~0.3 mbar; pressure was maintained while microwaves were applied to generate O_2 plasma for 15 s. The plasma cleaning process here also can activate radicals on the silica surface, which can form chemical bonds during the film deposition. After cleaning, the oxygen valve was closed and the vacuum in the chamber was restored to ~0.1 mbar. Meanwhile, liquid PDMS oil (4 ml) was heated at controlled temperatures to generate vapor, which is in thermal equilibrium with its liquid phase. Here, varied temperatures ranging from 50 °C to 130 °C were investigated for the impacts on film thickness and wetting properties. At the deposition process, microwaves were applied to activate air plasma in the vacuum chamber with the vapor fed through a needle valve. The plasma-aided chemical deposition was maintained for different periods of time, e.g. 5, 8, and 10 s.

2.4 Film characterizations

Static wetting properties of coating surfaces were evaluated through contact angles (CA) of distilled water using an optical contact angle measurement system (Dataphysics OCA-15EC). The contact angle measurement utilized probe liquids of droplet size ~2 μL placed at several locations on a sample surface. Dynamic wetting properties were determined via roll-off angle (RA) measurement of water droplet ~15 μL on a precise tilted platform. Surface topological structures and thickness of the deposited film were studied using scanning electron microscopy (SEM, FEI Quanta 400). The transmittance and reflectance spectra in visible region were recorded using a spectrometer (Blue-wave visible, StellarNet). The change of surface functional groups of the thin film was studied using Fourier transform infrared (FTIR) spectrometer (Bruker EQUINOX 55) operating in attenuated total reflectance (ATR) mode.

3. Results and discussion

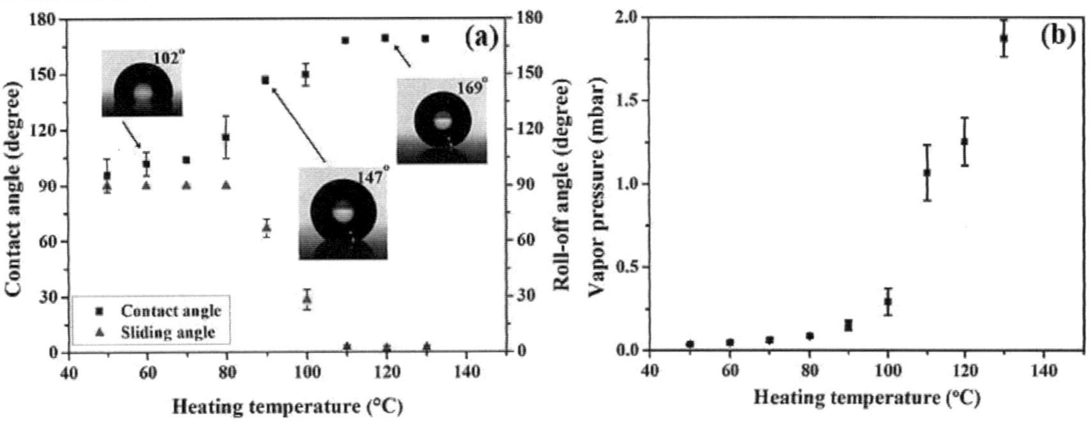

Figure 2. Contact angle and roll-off angle measurements of thin films deposited via MW-PECVD with various PDMS heating temperatures (a). Here, all plasma processes were maintained for 5 s. The change of PDMS vapor pressure versus heating temperature is shown in (b).

The effect of the PDMS heating temperature upon resultant water-repellent properties are unveiled through CA and RA values as shown in **Figure 2** (a). At low heating temperature regime, e.g. below 80 °C, the coated films show merely hydrophobic properties with contact angles residing in the range of 90-110°, which are similar to those obtained from smooth PDMS layers. Also water droplets stick on the surface without rolling. The absent evidence of roughness-amplified CA suggests that the deposited film is thin and smooth owing to low deposition rate caused by dilute PDMS molecules delivered into the reaction chamber. This is confirmed by the corresponding low vapor pressure at the same temperature range as shown in **Figure 2** (b). Furthermore, increasing heating temperature beyond 80° C significantly improves film CA and reduces RA values, indicating that surface roughness starts to play a role and the Cassie-Baxter non-wetting state is predominant. Results show excellent water-repellent properties with a maximum CA of 169° and a minimum RA of 3°, achieved with heating temperatures above 110° C. Albeit the ATM boiling point of the PDMS is greater than 140° C the vacuum environment helped to accelerate the vaporization rate and thus raises its vapor pressure.

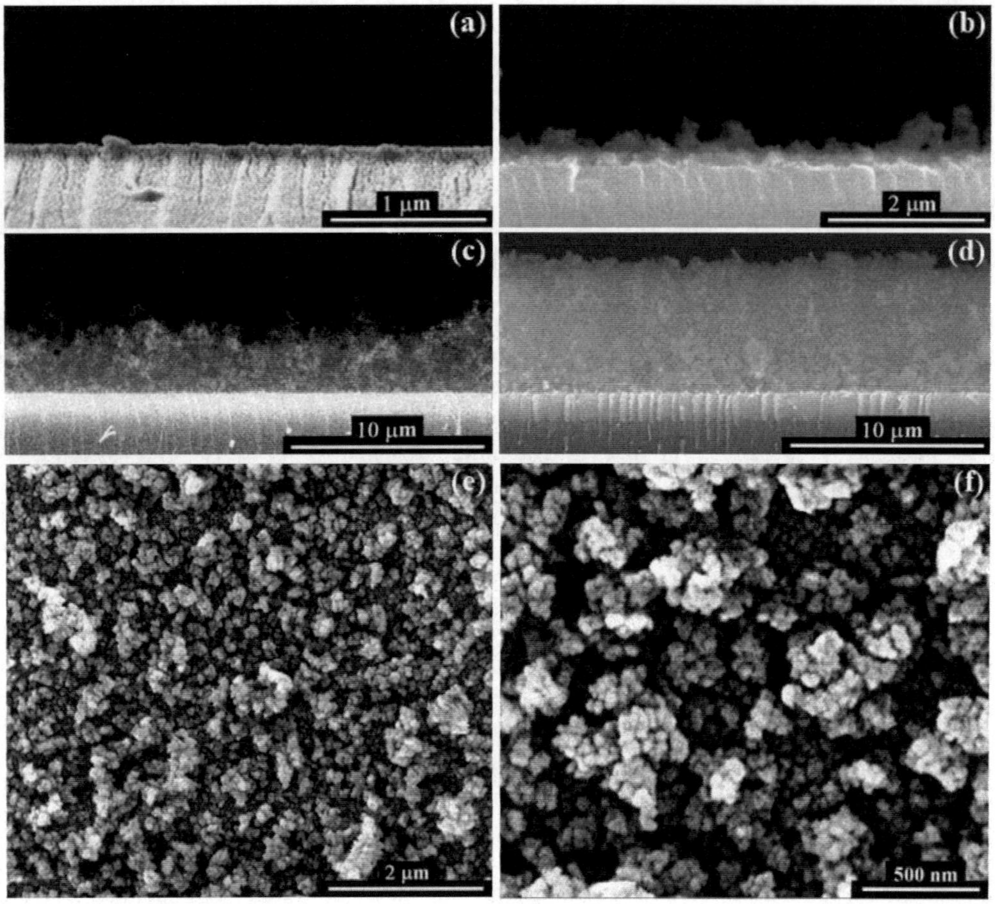

Figure 3. SEM micrographs show cross sections of a CVD film (a), and MW-PECVD films with deposition time of 5 s (a), 8 s (b) and 10 s (c) respectively. Plane views of the MW-PECVD films with deposition time of 5 s, (e) and (f), show highly hierarchical roughness.

We decided to maintain the material heating temperature at 120° C and investigated the influences of plasma processing duration upon the films structural, wetting, and optical properties. As displayed in **Figure 3**, SEM images reveal cross sections of the MW-PECVD films deposited for 5 s, 8 s, and 10 s. Their corresponding thicknesses are 0.34±0.09 μm, 3.5±0.4 μm, and 6.9±0.2 μm, respectively, averaged from multiple samples. Results indicate rapid growth of film thickness with increasing deposition time, inferring a very quick coating mechanism. It also can be observed a highly porous and irregular surface morphology for all plasma-deposited cases, in comparison to a dense and smooth layer obtained from the CVD counterpart. Top views of the SEM images also confirm random fractal-like surface features, with roughness sizes mainly in submicron and nanoscale, which are highly favorable for wetting-resistant phenomena.
For the optical properties of the MW-PECVD films, light transmission and reflection in visible spectrum were determined for samples with different coating times with the results presented in Figure **4**. The average transmission of a bare glass is ~ 89% with a hydrophilic surface CA of 34°. In contrast, all of the coated samples show excellent superhydrophobic properties with CA values near 170°. Overall, results are as expected, as increasing deposition time reduces the transmission due to more

absorption and scattering from films with thicker optical length. Of particular note is the sample with a deposition time of 5 s, where the transmission is even better than the bare glass sample. This is attributed to a couple of film properties. First, the thickness of the film is only 0.34±0.09 μm which greatly minimizes optical loss. Actually, this thickness result is consistent with previous work that suggests the optimized thickness for best coupling of transparent and superhydorphobic properties is below ~400 nm [4]. The other is the partial anti-reflection effect generated from the porous layer with nearly matched the refractive index [13]. Figure 4(b) show that the reflectance of all plasma-deposited films is lower than that of bare glass. In fact, thicker films show better anti-reflection behavior compared with thinner ones; however, their competing absorption and scattering grow faster. From results in Figure 4 it can deduced that the superhysdrophobic coating with good optical transparent can be actually fabricated just within 5 s. This unrivaled coating time is made possible through a high electron density provided by the MW plasma.

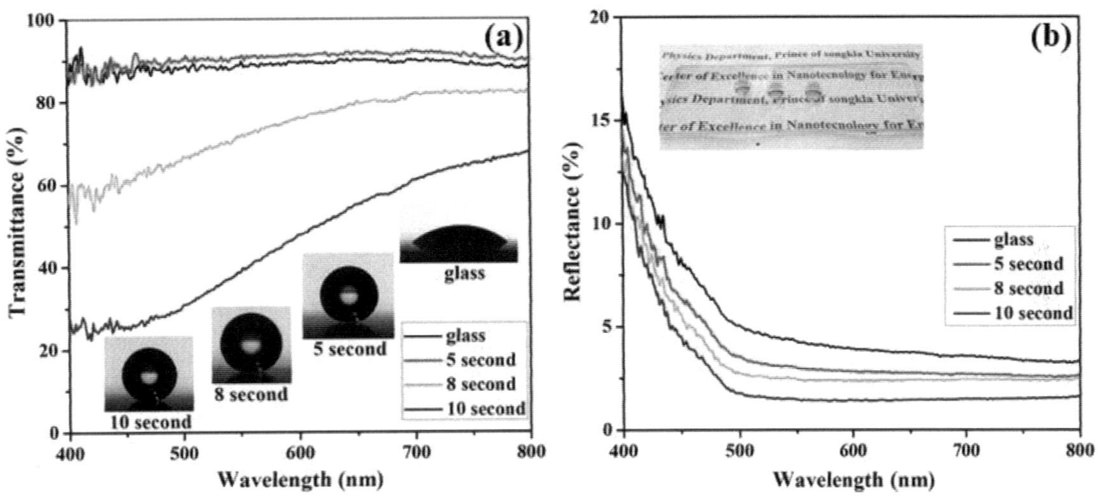

Figure 4. Transmission (a) and reflection (b) spectra of MW-PECVD films with different deposition times with their corresponding CA values included. The inset is a photograph showing water beads on a coated glass with high transparency.

The influence of MW plasma on the films molecular structure was investigated through infrared absorption technique, with two major changes of functional groups observed for a MW-PECVD film in comparison with one deposited via conventional CVD, as shown in Figure 5. Firstly, the broadening of the band centered around 1,050 cm^{-1} in the plasma-deposited case is suggesting an increasing degree of Si–O–Si network formation when compared to normal PDMS film, whose Si-O-Si characteristic spectrum generally consists of doublets at 1,085 cm^{-1} and 1,020 cm^{-1} [14, 15]. This particular spectrum broadening, as analogous to polyhedral oligomeric silsesquioxanes (POSS) structure, also indicates the existence of Si atoms, which bond to more than two O atoms that may involve in 2-D or 3-D crosslinks [16]. Another interesting point is the increase of the Si-O-Si/Si-CH_3 ratio for MW-PECVD films, which can be observed as the descent of the Si-CH_3 peaks at 780 cm^{-1}, 1,250 cm^{-1}, and 1,410 cm^{-1}, as well as Si-$(CH_3)_2$ and Si-$(CH_3)_3$ bands around 890-740 cm^{-1}. These FT-IR characteristics indicate a possible removal of methyl side groups due to the electron impact during the plasma process [17] followed by cross-linking and/or branching reactions forming complex network structures.

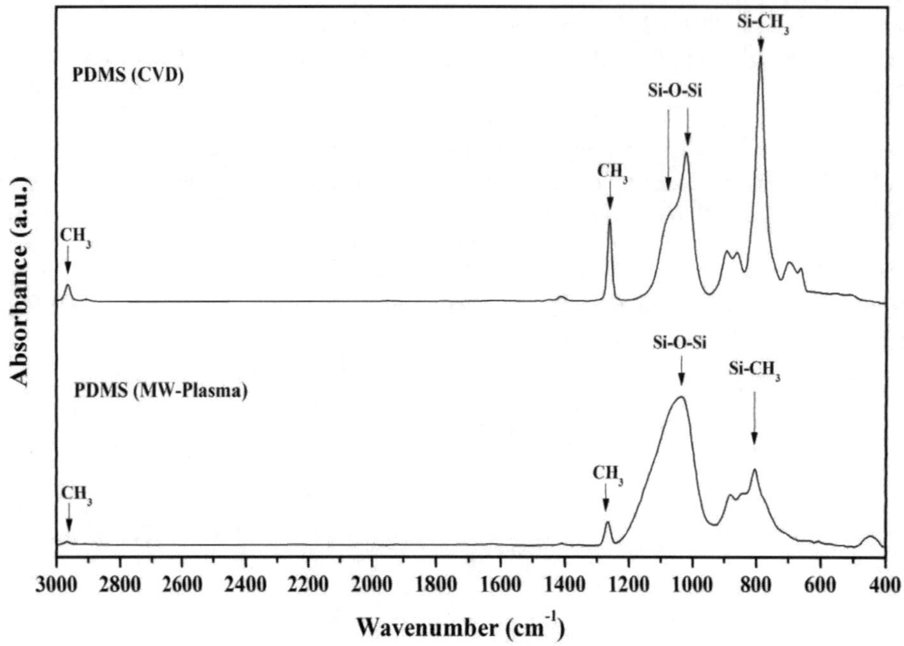

Figure 5. Comparison of FT-IR spectra of PDMS thin films deposited by (top) CVD technique and (bottom) MW-PECVD for 5 s.

The overall mechanism of the MW-PECVD process can be postulated based on information extracted from the abovementioned FT-IR results. As the reaction occurs in non-equilibrium MW plasma with high electron density and high power density, the vaporized PDMS substances are expected to suffer from molecular dissociation, excitation, and ionization caused by intense electron collision. The possible reaction schemes are proposed in Figure 6, where the electron impact results in PDMS chain scission producing smaller fragmented radicals. Also, the collision would create radicals at the polymer sidechain due to methyl abstraction process. This permits the formation branching or networked structures. Another competing chemical process taking place simultaneously is the plasma polymerization, which combines those dissociated radicals. Since the reactions are at relatively high energy and high entropy states, the subsequent structures are highly irregular with a high degree of roughness, favorable to superhydrophobic properties.

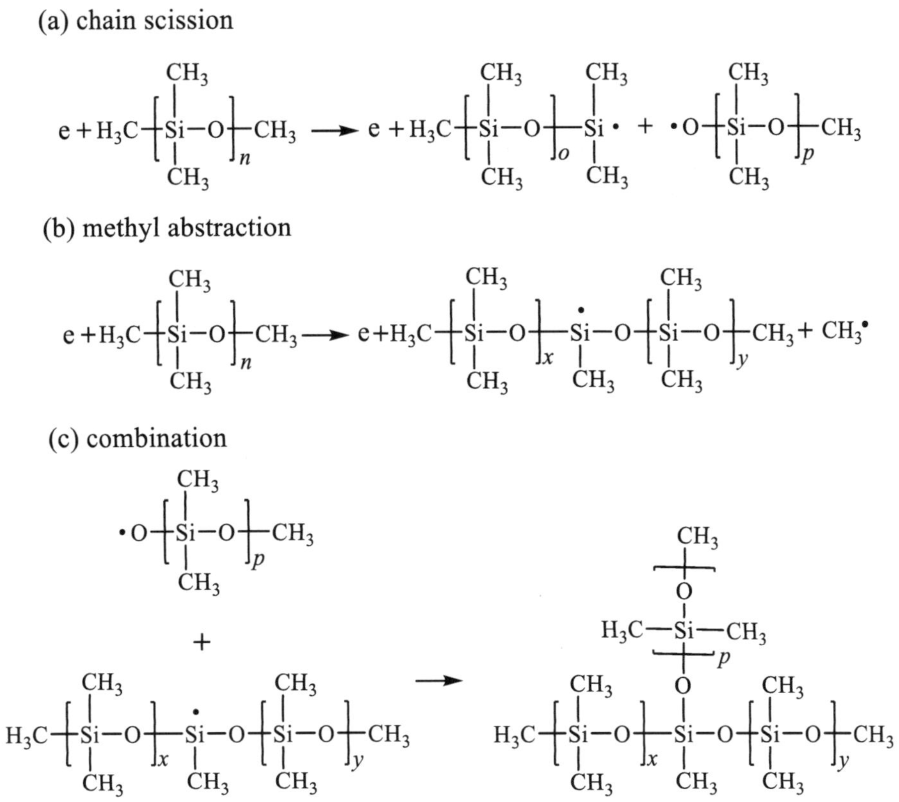

Figure 6. Possible chemical processes in MW plasma involving electron impact reactions, causing polymer chain fragmentation (a) and methyl abstraction (b). Those reactive components react and form more complex structures as shown in (c).

4. Conclusion

MW-PECVD technique has been demonstrated as efficient and environmental-friendly technique to produce transparent superhydrophobic surface within single coating step (excluding sample cleaning). PDMS liquid, used as a coating material, was vaporized and delivered into a MW-plasma chamber allowing electron impact processes to decompose the polymer chains into fragmented radicals. These highly reactive chemical species recombined via plasma polymerization, thus forming a disorder solid networked structure deposited on a glass substrate. The threshold of precursor vapor pressure to achieve multiscale roughness coating is above 0.25 mbar, which results in superhydrophobic properties. Furthermore, plasma processing time has shown a strong influence on film thickness and subsequent wetting and optical properties. With optimized plasma deposition time of merely 5 s, a superhydrophobic layer (CA ~170°, RA ~3°) with a thickness of ~340 nm can be fabricated. This ultrathin coating layer causes virtually no optical loss as the film also possesses anti-reflection properties which permit more light to pass through.

5. References

[1] Kota A K, Li Y, Mabry J M and Tuteja A 2012 Hierarchically Structured Superoleophobic Surfaces with Ultralow Contact Angle Hysteresis *Advanced Materials* **24** 5838-43

[2] Chunglok A, Muensit N and Daengngam C 2016 Extreme Wetting-Resistant Multiscale Nano-/Microstructured Surfaces for Viscoelastic Liquid Repellence *Journal of Nanomaterials* **2016** 13

[3] Schade H and Smith Z E 1985 Mie scattering and rough surfaces *Applied Optics* **24** 3221-6

[4] Cho K L, Liaw I I, Wu A H F and Lamb R N 2010 Influence of Roughness on a Transparent Superhydrophobic Coating *The Journal of Physical Chemistry C* **114** 11228-33

[5] Yang X, Gao P, Yang Z, Zhu J, Huang F and Ye J 2017 Optimizing ultrathin Ag films for high performance oxide-metal-oxide flexible transparent electrodes through surface energy modulation and template-stripping procedures **7** 44576

[6] Mahadik S A, Mahadik D B, Kavale M S, Parale V G, Wagh P B, Barshilia H C, Gupta S C, Hegde N D and Rao A V 2012 Thermally stable and transparent superhydrophobic sol–gel coatings by spray method **63** 580-6

[7] Karunakaran R G, Lu C-H, Zhang Z and Yang S 2011 Highly Transparent Superhydrophobic Surfaces from the Coassembly of Nanoparticles (≤100 nm) *Langmuir* **27** 4594-602

[8] Xu L, Karunakaran R G, Guo J and Yang S 2012 Transparent, Superhydrophobic Surfaces from One-Step Spin Coating of Hydrophobic Nanoparticles *ACS Applied Materials & Interfaces* **4** 1118-25

[9] Yoon S C a J S L a S B J a J G H 2013 Super-hydrophobic coatings with nano-size roughness prepared with simple PECVD method *Journal of Physics D: Applied Physics* **46** 315501

[10] Irzh A, Ghindes L and Gedanken A 2011 Rapid Deposition of Transparent Super-Hydrophobic Layers on Various Surfaces Using Microwave Plasma *ACS Applied Materials & Interfaces* **3** 4566-72

[11] Su X, Li H, Lai X, Zhang L, Liang T, Feng Y and Zeng X 2017 Polydimethylsiloxane-Based Superhydrophobic Surfaces on Steel Substrate: Fabrication, Reversibly Extreme Wettability and Oil–Water Separation *ACS Applied Materials & Interfaces* **9** 3131-41

[12] Szabó V D and Schlabach S 2014 Microwave Plasma Synthesis of Materials—From Physics and Chemistry to Nanoparticles: A Materials Scientist's Viewpoint *Inorganics* **2**

[13] Keshavarz Hedayati M and Elbahri M 2016 Antireflective Coatings: Conventional Stacking Layers and Ultrathin Plasmonic Metasurfaces, A Mini-Review *Materials* **9**

[14] Long D A 2004 Infrared and Raman characteristic group frequencies. Tables and charts George Socrates John Wiley and Sons, Ltd, Chichester, Third Edition, 2001. **35** 905-

[15] Suttor D, Kleebe H-J and Ziegler G 1997 Formation of Mullite from Filled Siloxanes **80** 2541-8

[16] Liu G, Hung W-S, Shen J, Li Q, Huang Y-H, Jin W, Lee K-R and Lai J-Y 2015 Mixed matrix membranes with molecular-interaction-driven tunable free volumes for efficient bio-fuel recovery *Journal of Materials Chemistry A* **3** 4510-21

[17] Pryce Lewis H G, Edell D J and Gleason K K 2000 Pulsed-PECVD Films from Hexamethylcyclotrisiloxane for Use as Insulating Biomaterials *Chemistry of Materials* **12** 3488-94

IOP Conf. Series: Materials Science and Engineering **311** (2018) 012016 doi:10.1088/1757-899X/311/1/012016

Evaluation of the residual stresses in 95wt%Al2O3-5wt% SiC wear protection coating using X-Ray diffraction technique

Adel K. Mahmoud[1], Zaid S. Hammoudi[2], and M.Sc. Student Samah Rasheed[3]

1 Professor in Mechanical Engineering Dept., University of Diyala-Iraq
2 Lecturer in Mechanical Engineering Dept., University of Diyala-Iraq
3 M.Sc. Student in Mechanical Engineering, Dept., University of Diyala-Iraq
E-mail:adel_alkayali@yahoo.com

Abstract. This paper aims to measuring the residual stresses practically in wear protection coatings using the sin2ψ method according to X-ray diffraction technique. The wear protection coatings used in this study was composite coating 95wt% Al2O3-5wt% SiC, while bond coat was AlNi alloy produced by using flame spraying technique on the mild steel substrate. The diffraction angle, 2θ, is measured experimentally and then the lattice spacing is calculated from the diffraction angle, and the known X-ray wavelength using Bragg's Law. Once the $d_{spacing}$ values are known, they can be plotted versus sin2ψ, (ψ is the tilt angle). In this paper, stress measurement of the samples that exhibit a linear behaviour as in the case of a homogenous isotropic sample in a biaxial stress state is included. The plot of $d_{spacing}$ versus sin2ψ is a straight line which slope is proportional to stress. On the other hand, the second set of samples showed oscillatory $d_{spacing}$ versus sin2ψ behaviour. The oscillatory behaviour indicates the presence of inhomogeneous stress distribution. In this case the X-ray elastic constants must be used instead of Young's modulus (E)and Poisson ratio (ν)values. These constants can be obtained from the literature for a given material and reflection combination. The value of the residual stresses for the present coating calculated was compressive stresses (-325.6758MPa).

1. Introduction

Growing use of thermal spray coatings, specially for high-temperature environmental resistance, needs assurance in coating durability, i.e. debonding, resistance to cracking, and spallation, both during appliance and in service. Residual stresses are acknowledged to play important roles in coating durability; i.e., tensile residual stresses commonly increase the susceptiveness to cracking and debonding. Various studies have been concerned to the measurement of residual stresses in coatings [1-4]. Residual stresses advance during cooling of a thermal spray coating because of the dissimilarity of thermal expansion coefficients of the coating and substrate. Depending on the relative consequences of the thermal expansion coefficients of the coating and substrate [5-7] residual stress can be either tensile or compressive Parameters that heavily affect the value of residual stresses are coating and substrate temperature during spray deposition and properties of the coating such as thickness, roughness and porosity. Experiments have demonstrated that residual stresses increase with coating thickness and deposition temperature [8]. Thermal barrier coatings (TBC) are the finest way to preserve components of gas turbine engines and the demand for this coatings is becoming more vital as higher temperature engines are being advanced [9-14]. Normally, the residual stresses of thermally sprayed coatings are made by diverse mechanisms and sources [15-17]. In a thermal spray procedure

with a high flame temperature, such like plasma spray, flame spray, or arc spray, entirely and partially molten particles striking onto the surface of the substrate, are became flat, solidified, and cooled down in a very small period of time (few microseconds). After their solidification and adhesion onto the surface of the substrate, the shrinkage of the splats can be stalled by substrate material or the underlying solidified coating material, which outcomes in tensile stresses which are called intrinsic, deposition, or quenching stress. Due to a very high temperature difference, a high theoretical residual stress in the order of up to 1 GPa can be made. However, Because of the many relaxation mechanisms, such as micro cracks, the sliding of the splats, plastic deformations, and material creep, the experimentally measured values are much lower (<100 MPa) [18]. X-ray diffraction was used as a corresponding technique; it can measure stress only in a thin surface layer, whereas the penetrating power of neutrons allows through-thickness stress profiling without any material exclusion [19].
This work aims to evaluation of the residual stresses in 95 wt% Al2O3-5wt% SiC wear protection coating coating using X-ray diffraction technique (XRD).

2. Materials and method

2.1. Materialsand parameters of the spraying processes

The coatings were applied by thermal spraying method (flame spraying) in air on the plain-carbon steel (AISI 1050), cylindrical substrate that's dimensions were 15 mm in diameter and 10 mm in height. The flame spraying system is designed and implemented in the welding laboratory of Mechanical Engineering Department, College of Engineering, University of Diyala, Iraq using spray gun the heat flame is produced by the burning of oxygen and acetylene, where the molten powder is carried out in the gas mixture and is attached to the surface to be coated by the high temperature of the torch which can raise to 3000 ° C. It is required to control the pressure of the gases to obtain the flame equal to the speed of the powder rush. The oxygen pressure should be adjusted according to the spray gun used no more than 4 bar and the acetylene pressure not more than 0.7 bar before spraying process. Two layers coatings were used in this work are the bond coat from (AlNi) alloy to reduce of thermal expansion coefficient between substrate and composite ceramic coating as a top coating layer Conditions of deposition process are listed in Table1.

Table1. Operating parameters during coating deposition process

Operating Parameters	Values
Oxygen pressure	4 bar
Acetylene pressure	0.7 bar
Distance	20 cm
Powder feed rate	$7\ cm^3$/min
Particle size	Mish (100-300)
Temperature substrate	$(350-400)^{0}C$

2.2.Residual stress analysis

2.2.1. X-ray stress evaluation. XRD-based residual stress measurements were made using standard ds_{pacing} vs. sin2ψ techniques using Shimadzu X-Ray Diffractometer type XRD-6000 and CrKα radiation. The $\sin^2\psi$ method [20, 21] was used to determine the residual stresses in this work, the change of a lattice plane distance (d spacing) of a phase, i.e., the peak shift of the corresponding reflection, was measured for tilt ψ-angles between 0° and 45°. To calculate the residual stresses the linear regression of the plot ($d_{spacing}$) versus $\sin^2\psi$ and the X-ray elastic constants. The coating and substrate physical properties (elastic modulus, Poisson's ratio, and coefficient of thermal expansion), thickness of the top coating, bond coat and substrate are shown in the table 2. The deposition temperature used in the present work during coating process and modeling for the top coating, bond coat and substrate was 850^oC.

IOP Conf. Series: Materials Science and Engineering **311** (2018) 012016 doi:10.1088/1757-899X/311/1/012016

Table 2. The coating and substrate physical properties [22-25].

	Substrate	Bond coating	Top coating
Young's modulus (Gpa)	200	105	65.78
Poisson ratio	0.33	0.315	0.231
Thickness	10 (mm)	200μm	500μm
Thermal expansion coefficient α, με K⁻	12.6	11.9 7.825	

3.Results and discussion

From Shimadzu X-Ray Diffractometer XRD-6000 chart, will be getting on the following values are shown in the Table 3.

Table 3. Showing 2Θ & ψ to 95%Al_2O_3& 5%SiC

2Θ	**ψ**
1.169565983	0
1.168720831	0.06698
1.168310272	0.25
1.166402633	0.5

By Brag Law (2d sinθ = nλ) may be calculated (d), where n=1, λ= 2.28970 A° and Θ (0,15, 30, 45) degree. From the figure1 may be calculated the linear slop of the plot $d_{spacing}$ versus $\sin^2\psi$

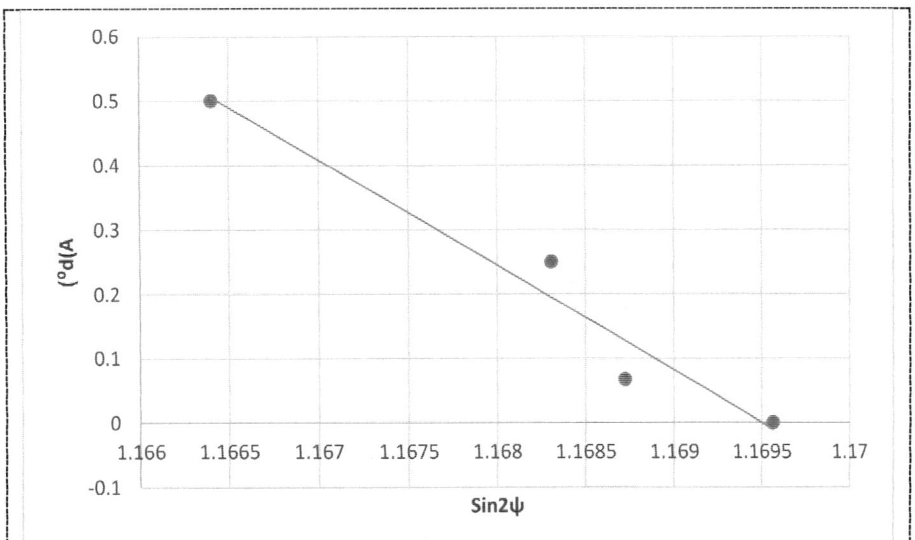

Figure 1. Relationship between (sin^2ψ) and $d_{spacing}$ (Å).

The stress can then be obtained from the following equation:

$$\sigma = (E/(1+v)) * 1/do\ (\partial d/(\partial\ 〖sin〗^2 d)) \qquad (1)$$

From figure1 the slop $\partial d/(\partial\ 〖\sin〗^2 d)) = -0.0058421$, $d_0 = 0.95856$.
from Eq. (1), the value of the residual stresses is: $\sigma = -325.6758$ MPa.

3.3.1. Quenchingstresses. Quenching stresses are of a level that corresponds to the values determined in the top surface of free standing coatings (Fig. 2a). They are always tensile and roughly constant through the deposit thickness. XRD data of quenching stresses, determined on the surface of different free-standing samples, are shown in Table 3.

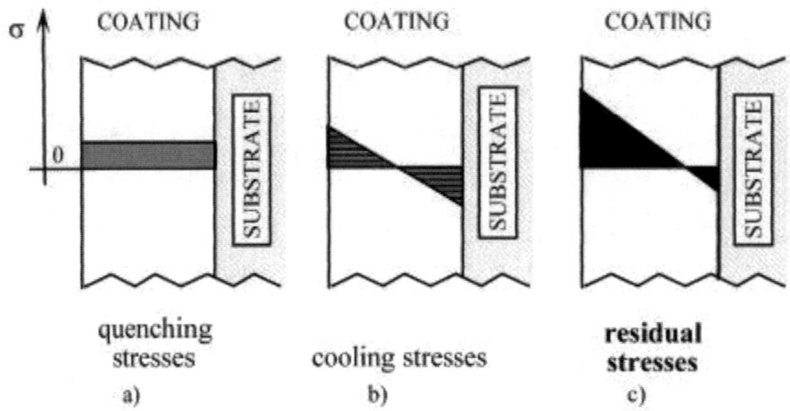

Figure2. Schematic explanation of residual stresses generation in thermal sprayed coatings [26].

3.3.2. Cooling stresses. The cooling stresses can be related to the effect of the coating-substrate connection by studying the residual stress relieving in free-standing coatings. For instance, the XRD data for 95wt% Al2O3-5wt% SiC flame spraying coating given in Table 3. The composite coating has a different behavior and the values remain constant, because of their brittleness, which induces mechanisms of strain accommodation inside the deposit. The value of the residual stresses is $\sigma = -325.6758$ MPa which was calculated using equation (1). This cooling stresses are due to the mismatch in thermal contraction between coating and substrate. They contribute to a stress gradient in dense coatings with a relatively high stiffness, because they are not affected by the release of quenching stress. In that case cooling component of the residual stress state increases with the coating thickness. The spalling of some thick coatings after spraying and during cooling to room temperature can be explained by the effects of the cooling stresses. So, interface with the substrate as a consequence of thermal mismatch strains. In fact, the difference of the stresses determined in the surface layer of these samples is opposite in sign to the cooling stresses calculated for the coating at the interface with the substrate. They are representative of the maximum bending stress at the surface side of the coatings while they are connected to the substrates. Cooling stresses, as responsible for a stress gradient in the coating thickness, explain the spalling of thick of materials that can be affected by their action. Industrial practice shows that such a phenomenon occurs, not during the spraying process, but after some time has elapsed, during the cooling to room temperature. This is explained by the bending moment due to the stress gradient, which would reach considerable magnitude in the thickest coatings. The residual stresses within the flame sprayed composite ceramic top coating are not distributed uniformly through the thickness, but change gradually, with the maximum in plane compressive stress at the interface, and diminish away from the interface, presenting very small value of residual stress on the free surface, this is confirmed by the study [27].

4. Conclusions

The X-ray diffraction technique has been demonstrated to be a powerful tool for evaluating and analyzing residual stress distributions in composite wear protection coating systems. In contrast to rather 'simple' single film on substrate systems, which are usually analyzed by means of the sin2ψ-method or related techniques, stress evaluation in complex coating systems requires the application of advanced methods that are different for the coating and the substrate.

In the present work may be drawn the following points:

1. The analysis of the in-plane stress was done in the individual layer of 95wt% Al2O3-5wt% SiC flame spraying composite coating.
2. The evaluation of the residual stresses is σ= -325.6758 Mpa which was calculated using the least-square-fitting the corresponding function to the experimental sin2ψ-data. This cooling stresses are due to the mismatch in thermal contraction between coating and substrate.
3. The residual stresses within the flame sprayed composite ceramic top coating are not distributed uniformly through the thickness, but change gradually, with the maximum in plane compressive stress at the interface, and diminish away from the interface, presenting very small value of residual stresses on the free surface.

5. References

[1] T.C. Totemeier, J.K. Wright, *Residual stress determination in thermally sprayed coatings-a comparison of curvature models and X-ray techniques, Surface & Coatings Technology*, 200 (2006), 3955 – 3962.

[2] Shubham Barnwal, B.C. Bissa, *Thermal barrier coating system and different processes to apply them- a review, International Journal of Innovative Research in Science, Engineering and Technology*, vol. 4, Issue 9, September 2015.

[3] J. Matejicek, S. Sampath, J. Dubsky, in: C.C. Berndt (Ed.), *Thermal Spray: A United Forum for Scientific and Technological Advances, ASM International,* Materials Park, OH, 1997, p. 855.

[4] S. Kuroda, Y. Tashiro, H. Yumoto, S.Taira, H. Fukanuma, S. Tobe, *Journal of Thermal Spray Technology,* 10 (2001) 367.

[5] O. Kesler, J. Matejicek, S. Sampath, S. Suresh, T. Gnaeupel-Herold, P.C. Brand c, H.J. Prask, *Measurement of residual stress in plasma-sprayed metallic, ceramic and composite coatings, Materials Science and Engineering*, A257 (1998) 215–224.

[6] Dietrich Mum, Matthias A. Sckuhr, and YingyuanYang,*Thermal Stresses in Ceramic-Metal Joints with an interlayer, J, Am. Ceram. Soc.*, 78 [21 285-90 (1995).

[7] S. R. Brown, I. G. Tuner, H. Reiter, *Residual stress measurement in thermal sprayed hydroxyapatite coatings', Journal of Materials Science: Materials in Medicine*, 5(1994), 756-759.

[8] R. Ghafouri-Azar, J. Mostaghimi, S. Chandra, *Modeling development of residual stresses in thermal spray coating, Computational Materials Science* 35 (2006) 13–26.

[9] C.U. Hardwicke, Y.C. Lau, *Advances in thermal spray coatings for gas turbines and energy generation: a review, J. Therm. Spray Technol.* 22 (5) (2013) 564–576.

[10] A.R. Krause, H.F. Garces, G. Dwivedi, A.L. Ortiz, S. Sampath, N.P. Padture, *Calcia-magnesia-alumino-silicate (CMAS)-induced degradation and failure of air plasma sprayed yttria-stabilized zirconia thermal barrier coatings, Acta Mater.* 105 (2016) 355-366.

[11] N.P. Padture, M. Gell, E.H. Jordan, *Thermal barrier coatings for gas-turbine engine applications, Sci.* 296 (2002) 280-284.

[12] A.G. Evans, D.R. Mumm, J.W. Hutchinson, G.H. Meier, F.S. Pettit, *Mechanisms controlling the durability of thermal barrier coatings, Prog. Mater. Sci.* 46 (2001) 505–553.

[13] C.R.C. Lima , S. Dosta , J.M. Guilemany , D.R. Clarke , *The application of photoluminescence piezo spectroscopy for residual stresses measurement in thermally sprayed TBCs, Surface & Coatings Technology* ,2016, SCT-21414.

[14] W. Fan, Y. Bai, Review of Suspension and Solution Precursor Plasma Sprayed Thermal Barrier Coatings, Ceramics International, 10 June 2016.

[15] J. Stokes and L. Looney, *Residual Stress in HVOF Thermally Sprayed Thick Deposits, Surf. Coat. Technol.,* 2004, 177-178, p 18-23.
[16] L. Pawlowski, The Science and Engineering of Thermal Spray Coatings, Wiley, New York, 2008.
[17] S. Sampath, X. Jiang, J. Matejicek, L. Prchlik, A. Kulkarni, and A. Vaidya, *Role of Thermal Spray Processing Method on the Microstructure, Residual Stress and Properties of Coatings: An Integrated Study for Ni-5wt %Al Bond Coats, Mater. Sci. Eng. A*, 2004, 364(1-2), p 216-231.
[18] W. Luo, U. Selvadurai, and W. Tillmann, *Effect of residual stress on the wear resistance of thermal spray coatings,* Submitted April 1, 2015; in revised form September 2, 2015.
[19] Olivera E. Kesler, *Residual stresses and properties of layered and graded coatings, Doctor of Science in Materials Science, Massachusetts Institute of Technology,* 1999.
[20] M.E. Fitzpatrick, A.T. Fry, P. Holdway, F.A. Kandil2, J. Shackleton and L. Suominen, *Determination of residual stresses by X-ray diffraction – Issue 2, National Physical Laboratory Teddington, Middlesex, United Kingdom, TW11 0LW*, September 2005.
[21] J. Pina, A. Dias, J.L. Lebrun, *Study by X-ray diffraction and mechanical analysis of the residual stress generation during thermal spraying, Materials Science and Engineering A*, 347 (2003) 21 /31.
[22] Touloukian YS, Kirby RK, Taylor RE, Desai PD.,*Thermophysical properties of matter thermal Expansion,* New York: Plenum Press; 1975.
[23] T.W. Clyne and S.C. Gill, *Residual stresses in thermal spraying coatings and their effect on the interfacial adhesion: A review of recent work, Journal of Thermal Spray Technology,*vol. 5(4) Dec. 1996.
[24] Xudong Sun, et al., *Intergranularparticleresidualstressstrengthening of Al_2O_3–SiCnanocomposite', Journal of American Ceramic Society*, 88 [6] 1536–1543 (2005).
[25] William F. Smith, JavadHashemi,*Foundations of Materials Science and Engineering,* Published by McGraw-Hill, Fourth Edition, 2006.
[26] J. Pina, A. Dias, J.L. Lebrun, *Study by X-ray diffraction and mechanical analysis of the residual stress generation during thermal spraying, Materials Science and Engineering A*347 (2003) 21/31.
[27] V. Teixeira, *Analysis of residual stresses in thermal barrier coatings, J. of Mat. Proc. Tech.* 92-93 (1999)209-216.

Acknowledgements
Authors wishing to acknowledge assistance and encouragement from colleagues in the Production Engineering Department & Metallurgy and Materials Engineering Department, University of Technology, Baghdad-Iraq, special work by technical staff in this University. Also acknowledge assistance from Mechanical Engineering Department, University of Diyala-Iraq.

IOP Conf. Series: Materials Science and Engineering **311** (2018) 012017 doi:10.1088/1757-899X/311/1/012017

Anti-corrosion properties of coatings with manganese compounds pigmentation

M Ziganshina[1] and E Nurislamova[2]

1 Department Head, Kazan National Research Technological University, Kazan, Russian Federation
2 Post graduate student, Kazan National Research Technological University, Kazan, Russian Federation
E-mail: zigmay4@mail.ru

Abstract. Work investigates properties of corrosion-resistant coatings based on organic-aqueous emulsion and pigmented compounds of manganese, obtained by ceramic method. It is found that the inclusion of synthesized pigments in the composition of the coating increases their ability to inhibit underfilm corrosion of steel.

Introduction

Currently, the exacerbation of environmental problems calls a particular attention to the toxicity of components in composition of paint and varnish material. One of the most pressing issues requiring immediate action is to replace toxic corrosion pigments from the composition of inhibiting type primers.

As a result, researches aimed at developing less toxic compounds that provide high protective properties of the primer coating became highly relevant. In recent years there was a number of works showing that one of the ways to reduce the toxicity of corrosion of primers is the replacement of chromium pigments on manganese compounds [1-4]. Manganese pigments described in these works were obtained by precipitation. The disadvantage of this method is the difficulty of cleaning of the derived pigments from water-soluble salts contained in the mother solution and an adverse effect on the protective properties of pigmented coatings. In this paper, as a corrosion pigments we take barium manganite-sulfate (BMS), calcium manganite (CM), barium manganate (BM) received in a ceramic way [5].

Table 1 lists the properties of the pigment compounds and zinc tetraoxychromate (ZTOH), known as the high-performance corrosion resistant pigment and is used in this work as a reference comparison.

Table 1. The properties of pigments used for filling

Pigment	Average oxidation state of manganese	Colour	pH of aqueous extract	Oil absorptio n 1 kind, g/100 g	Density, g/cm^3	Coverage rate, g/m^2	Content of substances soluble in water,%
BMS	4	Brown	8	14	3.7	26	0.7
CM	4	Grey	7	21	3.9	16	0.8

BM	5	Green	12	10	4.3	14	0.5
ZTOH	–	Yellow	7-8	30-35	3.41- 3.59	160-180	0.1-0.5

Protective properties of corrosion resistant coatings depends not only on the characteristics of pigment inhibitor included in its composition, but also on the nature of film-forming one. This work investigated the two classes of filmforming systems: organic- and water-borne. Determining factors when choosing the film-forming coatings are chemical resistant material, its coupling with the protected surface and insulating ability.

As the main factors contributing to corrosive processes behavior under coat, relates moisture, containing electrolytes, particular attention should be paid to the level of the insulating efficiency of the paint film. Also an important part of the protective properties of coatings is inhibiting the ability of compounds extracted from its components by water diffusing through the paint film. To evaluate the barrier and inhibiting properties of coatings based on synthesized pigments were used the results of values of electrical capacitance system of painted metal-electrolyte (C) and steady corrosion potential (E) of steel under cover, as well as a visual assessment of the state of specimens through the 1000 h exposure of 3% aqueous solution of sodium chloride on painted steel.

Experimental technique

Pigment pastes were obtained by dispersion of pigment parts of coating composition. We received degree of grinding 20-30 micron on device "Klin" using the laboratory beaded dispersant DISPERMAT VMA-GETZMANN GMBH D-51580 Reichshof. Coating deposited on previously degreased steel surface 0.8 kp in three layers by centrifuge method. Coating thickness through 7 days drying under natural conditions was 30 ± 3 micron. Capacitance of the system painted metal — electrolyte and electrochemical potential of metal under the coating in process of testing measured using emittance meter FLUKE PM6306 and pH-meter pH 340 respectively [6-8]. Potential measured relative to the silver chloride electrode and counted on the scale of normal hydrogen electrode.

Experimental results and discussion

Organic coatings based on film-forming systems

We assumed that as the basis of anti-corrosion primers for domestic purpose widely used alkyd oligomers, and in obtaining industrial primers applied epoxyamine binder. These affects choosing the type of organic aqueous film-forming coatings. Therefore, as objects of study were chosen alkyd binding (Spec 2311-023-45622449-2002) and epoxydiane oligomer (Spec-154 2225-05011907-97), with hardener polyethylenepolyamine (PEPA Spec 2413-357-00203447-99).

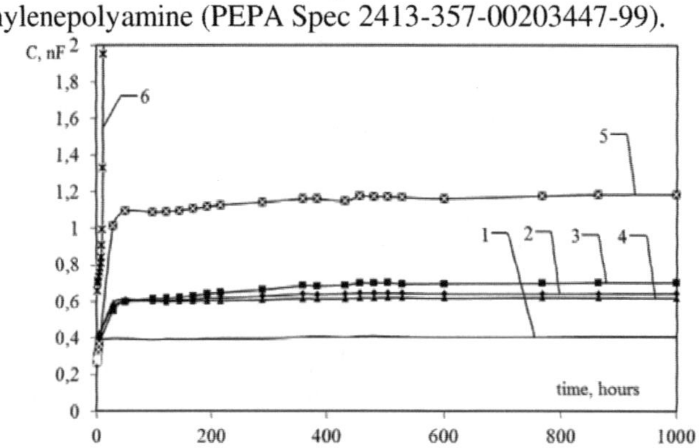

Figure 1. The dependence of electric capacitance values in the system painted metal - electrolyte from time of contact with coatings with 0.5 m aqueous solution of sodium chloride (for E-40 tracks with different levels of filling pigment HCMS% vol.)

Figure 1 presents the typical chronogram of values of capacitance that have the ascending branch due to diffusion of aqueous medium in the amount of paint film which goes into the sloping site, associated with the completion of the process of swelling. Violation of the integrity of paint coating, caused by exceeding the critical volumetric content of pigment, leads to the formation of direct contact of metal with electrolyte and, as a consequence of a sharp rise in the value of the capacity of the system due to the contribution of electrochemical capacitance (40% of the SWAPS curve in Fig. 1).

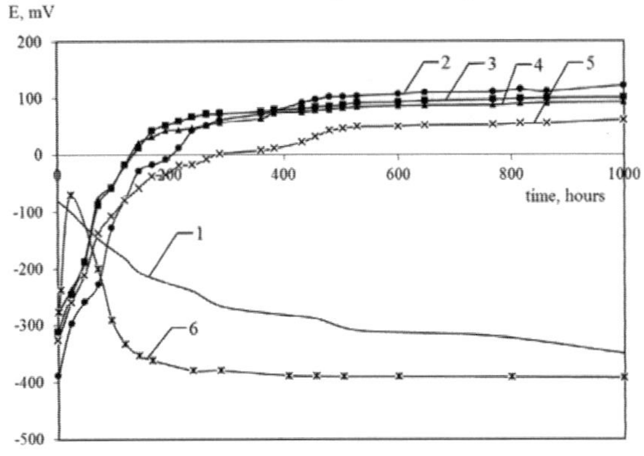

1 – 0; 2 – 20; 3 – 25; 4 – 30; 5 – 35; 6 – 40

Figure 2. Change of values of corrosion potential of system electrolyte - painted metal in time for epoxy compositions with various level of filling of SMES (% vol.)

The nature of chronopotentiometric curves, shown in Figure 2 indicates that the inclusion of manganiferous pigments in the composition of the paint film to a certain level, specific to each filming agent, greatly improves the potential of steel under the coating. The opposite occurs when the filling is above critical, in this case, painted steel capacity begins to decline. That is, the appearance of defects in the paint film causes the destruction of the passive film on the steel surface.

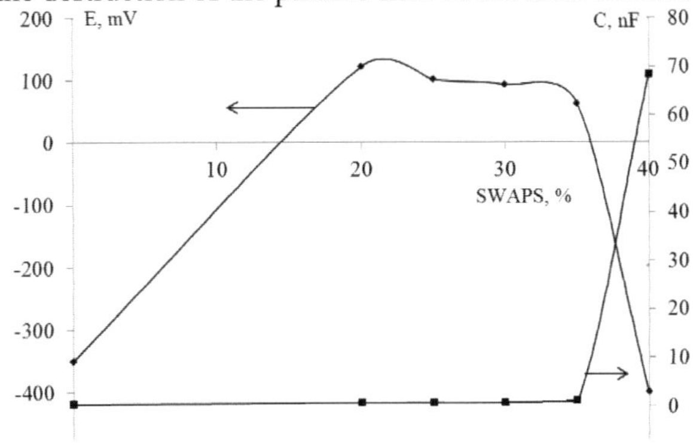

Figure 3. Changing the values of electrical capacitance and corrosion potential of system electrolyte - painted metal from filling level (for E-40 filled HCMs contact time of electrolyte with coating 1000 hours)

It is most clearly seen when presenting the data received in the form of established dependency values of capacitance and potential of volumetric content of pigment in the polymer. Such data for resin e-41 pigmented HCMS are presented in Fig. 3. The beginning of a sharp rise in capacity, corresponding to exceed the critical level filling occurs when filling is more than 35%. Decrease the insulating ability below a certain limit can not be offset by the inclusion in the cover even the most effective pigment-inhibitor.

Results of integrated assessment of coatings and metal substrate beneath it via 1000 h exposure on painted steel 3% aqueous solution of sodium chloride are presented in table 2.

Table 2. Results of integrated assessment

pigment	SWAPS,%	C, nF	E, mV	Area of bubbles, %	Area of corrosion %	Adhesion before testing	Adhesion after testing
alkyd binding							
None	0	5.95*	-360	5	10	2	2
	10	0.9	40				
BMS	20	1.1	260				
	30	1.3	210	0	0	1	1
	40	1.8	300				
	50	8.5	-300	2	10		2
	10	1.9	156				
	15	2.1	139	0	0	1	1
CM	20	2.6	127				
	25	3.2	-300	5			
	30	7.3*	-350	25	25	2	2
	10	0.8	114				
	20	1.0	102				
	30	1.2	54				
BM	35	1.3	23	0	0	1	1
	40	1.6	36				
	50	1.6	67				
	55	8*	-302		5	2	2
epoxydiane oligomer							
None	0	0.4	-350		10		2
BMS	10	0.65	112				
	20	0.71	102	0	0	1	1
	30	0.6	52				
	35	1.2	62				
	40	6.8	-400		20		2
CM	10	1.0	23				
	20	1.2	6	0	0	1	1
	30	1.2	-67				
	40	4.2	-350	25	20	2	2
BM	10	0.6	35				
	20	0.6	39	0	0	1	1
	30	0.7	27				
	40	1.1	-300	5		2	2

The data presented in the table allow you to identify possible filling limits upon receipt of corrosion primers based on synthesized pigments.

2. Coatings based on film-forming systems

The trend towards reducing the use of organic solvents in paint-and-lacquer materials production and application of water-dispersion paints is steadily increasing, expanding their product range.
Advantages of water-dispersion paint materials are obvious, they have no odor, dry quickly, easily applied to the surface. As a film-forming system for research has been chosen aqueous styrene-acrylate dispersion Lakroten E-241 (Spec 2241-031-51769913-2004), which, as preliminary studies have shown, suitable as film-forming the basis of anti-corrosion coatings.
The formed coatings were analyzed in the same way as the previous ones. The results may be seen in table 3.

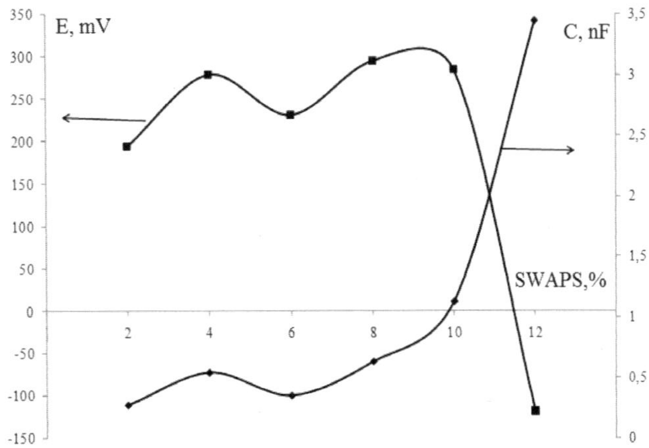

Figure 4. Changing the values of electrical capacitance and corrosion potential of system electrolyte — painted metal from filling level (for Lacroten-E241 filled with HCMs, contact time of electrolyte with coating 1000 hours)

Difference in values of boundary filling of water-dispersion varnish and water-dispersible filmforming systems (see figures 3 and 4) due to the fact that the latter are characterized by more high value relations of particle sizes of filmformer and pigment. In addition, the formation of the interphase boundary pigment — film binders in cover based on the pigmented emulsions is compounded by low mobility of macromolecules of filmformer, which may lead to a reduction in Interfacial adhesion and increase the defectiveness of the paint film.

Comparison of the results leads us to the conclusion that the nonpigmented coating has lower protective properties, since the corresponding values of corrosion capacity are at a lower level and monotonically decreasing. Integrated assessment results are presented in table 3.

Table 3. The results of integrated assessment

SWAPS,%	C, nF	E, mV	Bubbles, %	Corrosion area,%	Adhesion before testing	Adhesion after testing
BM						
2	0.262	284	0	0	1	1
4	0.538	117				
6	0.345	147				
8	0.7	373				
10	1.12	342				
12	3.44	-119		5		
BMS						
2	1.90	170	0	0	1	1
4	0.36	180				
6	0.50	213				
8	1.10	210				
10	1.50	170				
12	1.60	120				
14	8	-302		10		
CM						
2	0.26	200	0	0	1	1
4	0.29	211				
6	0.32	205				
8	0.39	214				

10	0.5	190				
12	0.6	178				
14	0.8	180				
16	1.2	170				
18	1.6	158				
20	4.5	-150		10		
Lakroten E-241						
0	3.963	-210	0	5	1	1

Summarizing the results of the comparative evaluation of protective properties of coatings, which are presented in tables 2 and 3, one can conclude that synthesized ceramic pigments HCMs MC and MB with SWAP level below the critical increase anticorrosion efficiency of coatings based on film-formers.

Conclusions

The influence of the manganese-containing anticorrosive pigments' content and nature on the protective properties of the coating has been studied. It has been shown that when filling below the critical level higher efficiency of anticorrosive coatings is observed. It allows to consider synthetic products alternative to low-toxic chromium-containing pigments.

References

[1] M.R. Ziganshina, M.S. Peshkova, S.N Stepin, V.E. Katnov: The effect of synthesis conditions on the properties of precipitated manganese-based anticorrosive pigments (*Lakokrasochnye Materialy i Ikh Primenenie*, 2005. No.10. - pp.34-36)

[2] M.R. Ziganshina, S.N. Stepin, M.S. Peshkova, A.A.Akhmadieva: Investigation of anticorrosive effectiveness of coprecipitative barium manganite-sulfate (*Lakokrasochnye Materialy i Ikh Primenenie*, 2005. No.3. - pp.16-19)

[3] M.R. Ziganshina S.N. Stepin, M.S. Peshkova, L.F. Dautova: Optical characterization of a white alkyd coating (*Lakokrasochnye Materialy i Ikh Primenenie*, 2004. No.8. - pp.3-6)

[4] V.A. Solodov, M.R. Ziganshina, E.A. Bayburina: Anticorrosive Efficiency of Coprecipitated Manganese Compounds IV (*International Journal of Applied Chemistry*, 2015. V. 11. No 5. - pp. 601-609)

[5] M.R. Ziganshina, S.N. Stepin, O .L. Afanasiyev: Evaluation of anticorrosive properties of manganese blue (*Lakokrasochnye Materialy i Ikh Primenenie*, 2007. No 9. - pp.19-22)

[6] S.N. Stepin: Method of valuation of critical volumetric content of pigments in the primer coatings (*Lakokrasochnye Materialy i Ikh Primenenie*, 1996. No.11. - pp. 12-15)

[7] S.N. Stepin, A.V. Vakhin, A.V. Sorokov, M.R. Ziganshina: Method to asses anticorrosive properties of pigments and pigmented coatings (*Lakokrasochnye Materialy i Ikh Primenenie*, 2000. No.1. - pp. 25-27)

[8] S.N. Stepin, A.V. Vakhin, M.R. Ziganshina, S.A. Karandashov: The study of anticorrosive properties of pigments (*Lakokrasochnye Materialy i Ikh Primenenie*, 2001. No.10. - pp. 3-7)

IOP Conf. Series: Materials Science and Engineering **311** (2018) 012018 doi:10.1088/1757-899X/311/1/012018

Predicting temperature drop rate of mass concrete during an initial cooling period using genetic programming

Santosh Bhattarai[1], Yihong Zhou[1, 3], Chunju Zhao[1], Huawei Zhou[2]
[1]College of Hydraulic and Environmental Engineering, China Three Gorges University, Yichang, 443002, China
[2] School of Water Resources and Hydroelectric Engineering, Wuhan University, Wuhan, China
[3]Correspondence: e-mail@ zyhwhu2003@163.com; Tel.: +86-13972605689

Abstract. Thermal cracking on concrete dams depends upon the rate at which the concrete is cooled (temperature drop rate per day) within an initial cooling period during the construction phase. Thus, in order to control the thermal cracking of such structure, temperature development due to heat of hydration of cement should be dropped at suitable rate. In this study, an attempt have been made to formulate the relation between cooling rate of mass concrete with passage of time (age of concrete) and water cooling parameters: flow rate and inlet temperature of cooling water. Data measured at summer season (April- August from 2009 to 2012) from recently constructed high concrete dam were used to derive a prediction model with the help of Genetic Programming (GP) software "Eureqa". Coefficient of Determination (R) and Mean Square Error (MSE) were used to evaluate the performance of the model. The value of R and MSE is 0.8855 and 0.002961 respectively. Sensitivity analysis was performed to evaluate the relative impact on the target parameter due to input parameters. Further, testing the proposed model with an independent dataset those not included during analysis, results obtained from the proposed GP model are close enough to the real field data.

Keywords: Mass concrete; Eureqa; prediction; genetic programming; temperature drop rate

1. Introduction

Mass concrete plays an important role in modern construction, especially in hydraulic and hydroelectric construction. For example, in China more than 10 million m^3 mass concrete are poured every year in hydraulic and hydroelectric engineering. Besides, the structure of harbor engineering and foundations of heavy machines are often built with mass concrete. In such structures, heat of hydration of cement increases the internal temperature of concrete due to its exothermic chemical reaction and induces thermal gap induces between inside and outside of structure leading to tensile stress which ultimately results in thermal cracking of such structures almost at early age [1, 2]. In order to reduce/control the maximum internal temperature of concrete and to fasten the cooling process, chilled water is circulated through the interconnected cooling pipes embedded into the concrete during construction [3, 4]. This technique was first studied in early 1930's by USA Bureau of reclamation in the design of Hoover Dam [5].

Thermal cracking in mass concrete structure at early age depends upon the rate of which the concrete is cooled (temperature drop rate per day, $\Delta t\ ^0C/day$) at an initial cooling period. Thus, temperature of mass concrete should be dropped at suitable rate to control the cracking of mass concrete during the construction phase. There are lots of factors to be considered to prevent thermal induced cracking of mass concrete at an early age. Reducing the maximum internal temperature of concrete, adjustment of suitable combination of water cooling parameters: flow rate (q_w), inlet temperature of cooling water (T_w)

and controlling the rate at which the concrete cools (temperature drop rate), are some of them. Further, Δt depends upon many parameters like, thermal properties of concrete, spacing of pipe, q_w, T_w, age of concrete (A_c), construction season (Summer or Winter) and so on. Relationships between these parameters are non-linear in pattern, complicated and not well understood yet. Due to the lack of simple and practical formula, in recent construction of concrete dams, T_w and q_w are adjusted based on numerical simulation and engineering experience to control the temperature of concrete.

A numbers of researches emphases on controlling temperature of mass concrete using numerical simulation [6-9]. Further, most recent researches conducted for determining the temperature field of concrete dam at construction phase are based on Finite Element Methods (FEM) are [10, 11], composite element method [12] and heat fluid coupling method [13]. Thermal crack analysis during pipe cooling was simulated using Particle Flow Code method [14]. However, very few researches have investigated the cooling rate of mass concrete during an initial cooling period at the construction phase of massive concrete structures. EM 1110-2-2201(1994) suggests not to drop the temperature of mass concrete during more than ½ to 1 ^{0}F per day during an initial cooling period [15]. ACI 207.4R (1993) reported the cooling rates, in degrees per day, for latter period should be lower than the permitted during initial periods because of the higher modulus of the elasticity at later ages [16]. Also this report suggests, after the concrete reached its peak temperature, cooling should be continued for a period of 1 to 2 weeks at a rate such that the concrete temperature drop generally not exceed 1^{0}F (0.6 ^{0}C) per day (the maximum rate that does not exceed the early age tensile strain and creep). For some condition, cooling rate of 2 ^{0}F (1 ^{0}C) per day can be accepted for a short period of time [16]. Nannan Shi et al. (2014) investigated; lower cooling rates can reduce the probability of concrete cracking [17]. Further, temperature control of mass concrete during the construction phase of concrete dams seems more complicated due to the varying properties of concrete with the passage of time (A_c). Moreover, construction time plays an important role in any project; thus, in order to complete the project within the time frame, it is often needed to pour the concrete in a high temperature summer seasons. If an effective temperature control measure is not taken for concrete poured at higher summer season undesirable and unavoidable cracks will be the result [18].

Therefore, in order to prevent the cracks during the construction phase of concrete dams, it is very important to understand the relationship between the parameters involved in the process. In the present study, attempts have been made to develop a robust prediction model to predict Δt at an initial cooling period during the construction phase of concrete dams. Data measured at summer construction season (April- August from 2009 to 2012) are used to formulate the relationship. For this, T_w, coefficient of pipe cooling (p_1) which is dependent q_w and A_c are taken as the input variables whereas values of Δt is taken as the target variable.

2. Materials

2.1. Data Source

In order to formulate the relationship to determine Δt, data were taken from the project named "Xiluodu high concrete arch dam (285.5 m high) which was recently constructed and located in the lower reach of the Jingsha River, Yunnan Province, in southwest China [19]. During the construction of the project, an optical fiber (shown in Figure 1) was embedded in concrete to monitor the temperature of concrete. Data's from monolith 15 and monolith 16 during the construction at summer season (April- August from 2009 to 2012) were chosen. Three types of concrete namely Concrete Type A, Concrete Type B and Concrete Type C were used while constructing the dam. In this study, Concrete Type A was used for formulating and verification of the developed model. Thermal properties of concrete (Concrete Type A) as: thermal conductivity, diffusivity, specific heat and density and water cooling data as: diameter of water cooling pipe, conductivity of pipe, length of cooling pipe and spacing of cooling pipe (H: V) as 1.5m*1.5m and lift height (1.5 m and 3m) were taken as the real situation of the research project. . Thermal properties of concrete are listed in Table 1.

Figure 1. On-site Schematic layout of optical fiber

Table 1. Thermal Properties of Concrete

Density(δ)Kg/m^3	Thermal Conductivity (λ) Kj/(m h ^{0}C)	Specific Heat (C) Kj/(kg ^{0}C)	Diffusivity (a) m^2/day
2592.58	7.35	1.08	0.063

2.2. Data Analysis and Preparation

Concrete gain its peak temperature after a few days to week of placement. After the concrete reaches its maximum/initial value, dropping of temperature from this value to the target value (design value) at an early age (around 30 days) is fully controlled by circulating water through the cooling pipes embedded in the concrete during construction. The time period that takes dropping the maximum/initial temperature to target temperature at early stage is known as initial cooling stage/period. The overall cooling process used during the construction phase is shown in **Figure 2**.

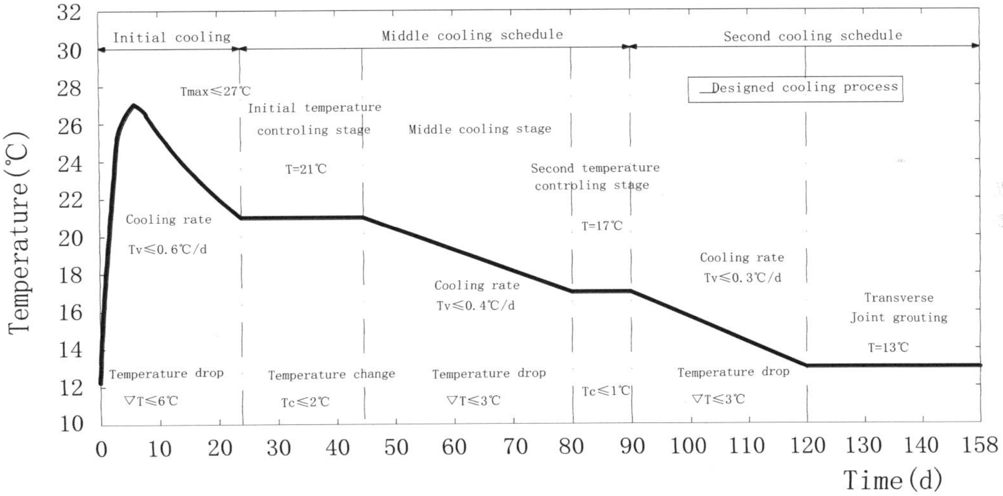

Figure 2: Overall Cooling Process of Xiluodu High Concrete Dam

To develop the prediction model for temperature drop rate, concrete temperature data (recorded from optical fiber at different time interval within a day for each lift taken in this study) and water cooling data (measured at almost the same time of concrete temperature measurement) within initial cooling stage are utilized in this study. Concrete temperature data are taken at the time lag of almost 12 - 24 hours between two consecutive days. For some cases, concrete temperature data of latter days are greater than previous days, in such cases, the value of Δt is negative and those negative values of Δt has been removed during developing the model. The set of data for Δt before building the model is prepared as follows:

$$\Delta t_i = t_{i-1} - t_i \quad (1)$$

Where,

$\Delta t = \Delta t_i$= temperature difference of mass concrete of [(i-1th) - ith] day in ^{0}C/day
t_{i-1} = temperature of concrete at ith-1 day in ^{0}C
t_i = temperature of concrete at ith day in ^{0}C and

i=2, n: n is total numbers of days at initial cooling period for individual lifts.

The relation given in equation (1) is demonstrated by the following example:
Suppose, for lift number 3, initial cooling period is 21 days ranging from date 2009/4/1: 8:00 PM to 2009/4/22: 8:00 PM. Concrete temperature at date 2009/4/1: 8:00 PM and 2009/4/2: 8:00PM is 25.6 ^{0}C and 25.2 ^{0}C respectively, then the value of Δt for day 2009/4/2: 8:00 PM is 25.6 - 25.2= 0.4 ^{0}C/day. For preceding days up to initial cooling period, the same concept applies. In this example i =2, n, 2 denotes the latter day 2009/4/2: 8:00 PM and n denotes the last date of initial cooling period, i.e., 2009/4/22: 8:00 PM. Similarly, for other lifts Δt is calculated in the same manner.
Further, coefficient of pipe cooling, p_1 (which depends upon q_w) can be derived from the following relationship, Zhu [20]. The rate of flow of cooling water is taken almost at the same time of calculating Δt.

$$p_1 = k_1 (ga / D^2)^s \tag{2}$$

Where,

$$k_1 = 2.08 - 1.17\xi + 0.256\xi^2 \tag{3}$$

$$s = 0.971 + 0.1485\xi - 0.0445\xi^2 \tag{4}$$

$$\xi = \frac{\lambda L}{c_w \rho_w q_w} \tag{5}$$

$$g = 1.67 \exp\{-0.0628[\frac{b}{c}(\frac{c}{r_o})^{\eta} - 20]^{0.48}\} \tag{6}$$

In which, g is a coefficient to consider the influence of b/c and the material of pipe

$$b = 0.5836 * \sqrt{(S1*S2)} \tag{7}$$

Where, *a* is the thermal diffusivity of concrete (m^2/day); *b* is the outer radius of concrete cylinder (m); *c* is the inner radius of concrete cylinder (m); c_w is the specific heat of cooling water (kJ/Kg ^{0}c); CD is the Construction Date ; D is the diameter of concrete cylinder (m); *L* is the length of pipe (m); r_0 is the inner radius of non-metal cooling pipe (m); S_1 is the horizontal spacing between cooling pipes (m); S_2 is the vertical spacing between cooling pipes; λ is the coefficient of thermal conductivity of concrete (kJ/m h ^{0}c); λ1 is the coefficient of thermal conductivity of non-metal cooling pipe (kJ/m h ^{0}c); ρ_w is the density of cooling water (kg/m^3); and η = λ/λ1

Calculation for A_c is demonstrated by the following example:
Suppose, for lift number 3, initial cooling period is 21 days ranging from date 2009/4/1: 8:00PM to 2009/4/22: 8:00PM. If the date of construction of lift 3 is 2009/3/25: 5:45: AM, then the age of concrete up to date 2009/4/1: 8:00PM is 2009/4/1:8:00PM - 2009/3/25: 5:45: AM = 7.59375 days. Likewise, the age of concrete for each day up to initial cooling period of lift 3 can be calculated for the exact time of calculating Δt. Likewise, for other lifts, age of concrete is calculated on the same principle. To formulate the GP model, descriptive statistics of the input data are tabulated in **Table 2**.

Table 2: Descriptive statistics of input data

	Input			Target
Rank	p_1	T_w (^{0}C)	A_c(days)	Δt (^{0}C/day)

Mean	0.0358513	11.95405	18.32993	0.221986
Median	0.035000	14.00000	17.71002	0.21150
STDEV*	0.006295	2.79579	6.68767	0.11293
Variance	3.9635E-05	7.81649	44.72497	0.01275
Maximum	0.046000	15.30000	31.11875	0.46700
Minimum	0.023000	6.00000	6.70104	0.00900

*=Standard Deviation

3. Methods

3.1. Genetic Programming, Eureqa and Automated Solution Seeking

Recently, soft computing technique (like artificial neural networks (ANNs) and fuzzy neural network systems),which are considered as strong machine learning techniques have been gaining popularity for solving complex problems in the field of civil engineering [21-23]. Genetic Programming (GP), which was first introduced by John Koza in (1992) [24] is another branch of machine learning method, which automatically generates computer programs based on the rule Darwinian natural selection and biologically inspired operations to solve the user-defined task. Genetic programming (as an extension of genetic algorithm) evolves a series of computer programs (semi complex mathematical equations) instead of data to solve a complex non-linear problem and can be suitable for real problem [25]. GP uses an evolutionary algorithm in order to optimize the computer programs through expression tree according to fitness function.

Eureqa® software package sometimes can be called as robot scientist, developed by Dr. Hod Lipson [26] is fairly new, publically available product from Cornell Creative Machines Lab [27] is a symbolic regression tool for automated numerical regression methods, optimization, detecting equations and hidden mathematical relationships in raw data and is based on GP. Eureqa has been applied for solving some problems in civil engineering field [25, 28-29]. Correlation coefficient (R) and Mean Square Error (MSE) were used to evaluate the performance of each model. In Eureqa, each variable values can be assigned to single rows and searches are specified by writing a search function. A solution fit plot against predicted and actual data, list of candidate function ranked by fitness (error/complexity), a plot of solution respective to their error size; residual error plot and a plot of different fitting statistics of the generated solutions can be obtained as output in Eureqa [28].

3.2. Development of model using GP

To get the suitable GP model, basic arithmetic operators (+, -, *, /), trigonometric operator (sin, cos) and some basic exponential functions (exponential, natural logarithm, square root, factorial and power) were utilized in this study. Un-normalized data of the individual variables has been normalized during building the model. The GP software identified un-normalized data of the individual variables. And those identified un-normalized data's from GP software for each variables is normalized by algorithm "subtracted by mean and divided by standard deviation". 145 numbers of rows of each input and an output variable were gathered. 51 numbers of data were chosen for training, 24 numbers of data were chosen for validation and 70 numbers of data were chosen for testing of the proposed model (chosen randomly). Following function is used to obtain the hidden relationship between Δt and the influencing variables:

$$\Delta t = f(p_1, T_w, A_c) \quad (8)$$

4. Results and Discussions

4.1. GP-Based Formulation for Δt

The GP approach was employed in this study to predict Δt of mass concrete at an initial cooling period during the construction phase of concrete dam. Final model is established by comparing the output from the developed model with the experimental data. The best two models evaluated from GP are shown in **Table 3**. The value of R close to 1 and low value of MSE indicates the data were more fitted. The value of R given in **Table 3** is for validation dataset. According to the performance evaluation criteria, model no 1 is the comprising model among others.

Table 3: Different GP Based Prediction Models

Rank	MSE	R	Expressions
1	0.002961	0.8855	$\Delta t = \dfrac{0.755 + 160p_1}{5.73 + A_c + 1.24T_w + \sin(2.08A_c)(0.00101A_c^{\,3} - 1.25) + 0.363A_c\sin(2.53\sin(4.76e4A_c)) - 160p_1}$
2	0.002971	0.8848	$\Delta t = \dfrac{0.63 + 160p_1}{5.73 + A_c + 1.19T_w + \sin(2.08A_c)(0.000999A_c^{\,3} - 1.25) + 0.358A_c\sin(2.54\sin(4.76e4A_c)) - 160p_1}$

Due to the lack of previously developed rational models to predict Δt at an initial cooling period during the construction phase of concrete dam, it is not possible to conduct a comparative study of the results obtained from this study to those of previous studies. According to Smith [30], when the model gives |R|>0.8, a strong correlation exists between the predicted and measured values. As can be seen from **Table 3**, the entire model has |R|>0.8, which reveals that the proposed model has a good predictive ability.

4.2. Performance of the Model

In order to determine the prediction capability of the proposed model, comparisons were made between the predicted values of Δt from GP model with real Δt that were not included during an analysis by plotting the graph as shown in **Figure 3**.

It is obvious from **Figure 3**, during testing phase the predicted and real Δt were strongly correlated with a linear relationship with R^2/R of 0.7202/0.8486 from the proposed model. Beside validation, performance of the derived model is verified by comparing the prediction output from the derived model with the real field data those were not included during the analysis as shown in **Figure 4**. Comparison (shown in **Figure 4**) was made between the predicted Δt and real Δt (data available from the same research project having different value of qw and T_w per day). The calculated results from proposed model shows a pretty good agreement with the real field data for summer month from April to August for different year which indicates that the proposed model is obvious.

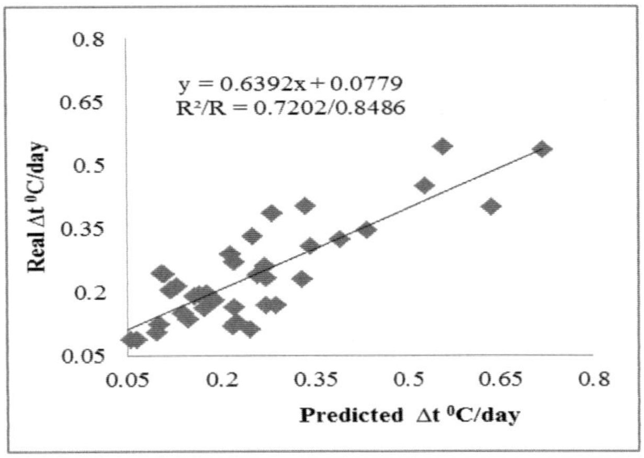

Figure 3: Comparison between real Δt and predicted Δt during the testing phase

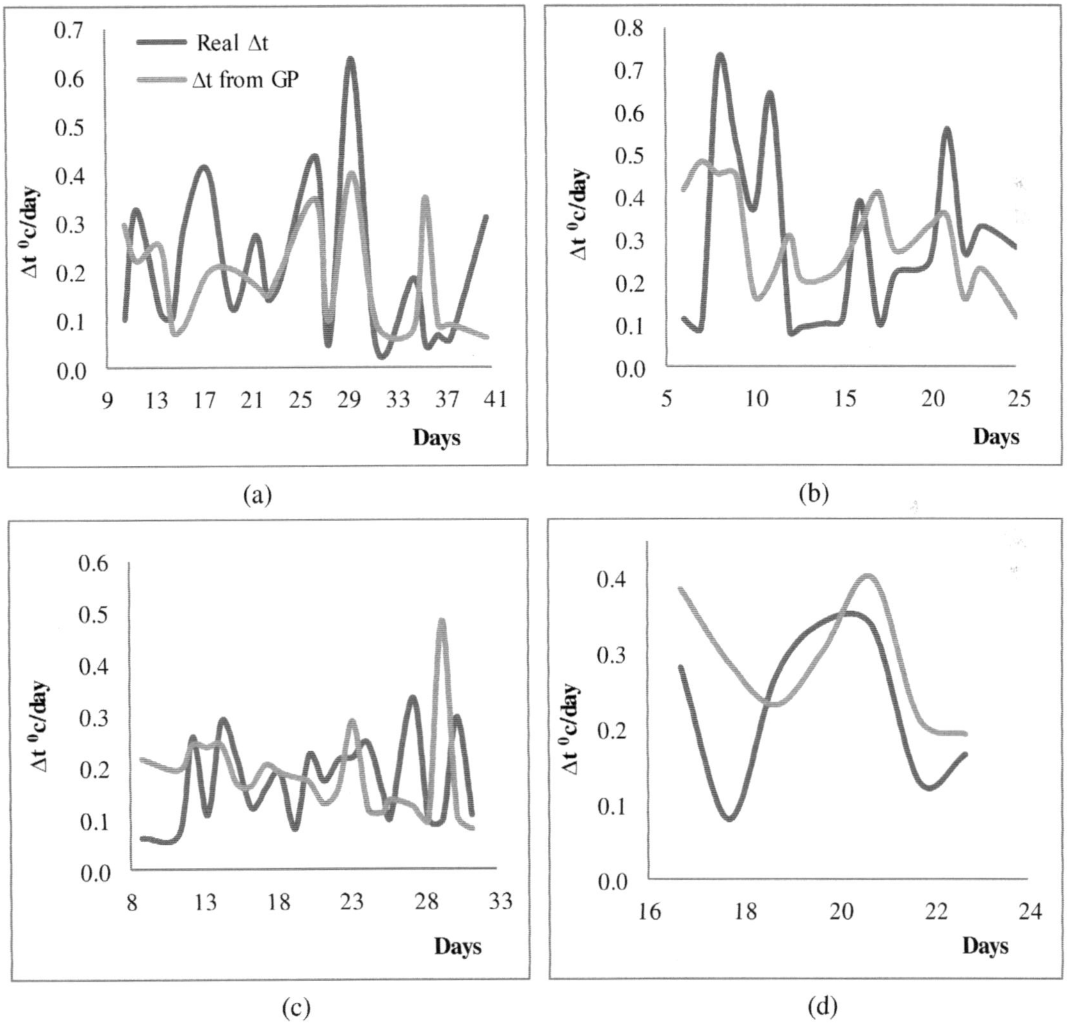

Figure 4: Comparative study of result from GP-model with real field data: (a) Lift 74 of monolith 15 (CD: 2012/4/17), (b) Lift 5 of monolith 15 (CD: 2009/5/9), (c) Lift 72 of monolith 16 (CD: 2012/6/16), (d) Lift 85 of monolith 15 (CD: 2012/8/30)

Note: Green and Grey line used in Figure 4b -Figure 4d is the same used in Figure 4a

4.3. Sensitivity Analysis

Sensitivity analysis was performed to evaluate the relative impact on the target variables due to input variables. For a given model in the form z=f(x, y,.....), sensitivity is expressed as follows:

$$Sensitivity = \overline{\left| \frac{\partial_z}{\partial_x} \right|} * \frac{\sigma(x)}{\sigma(z)} \tag{9}$$

Where: δ= partial derivate operator, σ(x) = standard deviation of x in the input data, σ(z) = standard deviation of z [27].

The term percent positive is defined as the percent of data in which the partial derivative of the target value with respect to the ith input is greater than zero. This number shows the possibility that increasing the specified input parameter would increase the target value in the model and the same concept applies for a negative value of the aforementioned derivative term known as percent negative. Further, positive magnitude is the number that denotes generally how big the positive impact is, when increases in this variable lead to increases in the target variable and same concept applies for a negative magnitude [27]. A summary of sensitivity, percent positive, positive magnitude, percent negative and negative magnitude values for the GP-based model is shown in **Table 4**.

Table 4: Sensitivity analysis of the GP model

Input Variable*	Sensitivity	Percent Positive	Positive Magnitude	Percent Negative	Negative Magnitude
p_1	0.37518	100	0.37518	0	0
T_w	0.28847	0	0	100	0.28847
A_c	36.536	41	38.882	59	34.898

*As defined in expressions in **Table 3**.

Sensitivity analysis result shown in **Table 4** indicates, input parameters p_1 and T_w are possess high positive and negative impact to the target parameter respectively, whereas parameter A_c has percent negative of 59% ≈ 60%, thus it is supposed to have negative impact on the target parameter.

The proposed GP model developed in this study was formulated from the data available from single dam constructed at summer season from April to August. Therefore, the model derived from this study can be used in preliminary design stages rather than in final decision making. Although, the derived model holds good agreement within the range of data given in **Table 2**, the model should be used carefully for prediction outside the range of parameter taken in this study. This proposed model could be improved to make a precise prediction over a wider range considering different pipe spacing and initial cooling duration, if the data could be made available from other similar projects. This research is expected to be helpful where high concrete dams are anticipated to build in near future.

5. Conclusion

The GP approach was employed in this study for predicting Δt of mass concrete at an initial cooling period during the construction phase of high concrete dam. This model will be capable to sort out the complication of adjusting the water cooling parameters (q_w and T_w) during the construction phase of concrete dam in order to drop the maximum temperature of concrete. The developed model has the R value of 0.8855(which is greater than suggested good fit |R|>0.8) and significantly low MSE (0.002961), which indicates the proposed model have good predictive ability. Beside validation, testing the model applicability with an independent dataset (those not included during analysis), proposed model is capable of generalizing the input and output variables with reasonably good predictions. Sensitivity analysis results clarifies that input parameters: p_1 possess positive impact and T_w is sensitive in-terms of negative impact on the target variable whereas parameter A_c has percent negative of 59% ≈ 60% which is supposed to be have negative impact on the target variable. Using the derived model, required Δt during initial cooling period at construction phase of concrete dam can be easily calculated using the variable taken in this study, which will be beneficial in-terms of time saving for sophisticated laboratory experiment.

Conflicts of Interest

The authors declare no conflict of interests regarding the publication of this paper.

Acknowledgments

This research was supported by National Natural Science Foundation of China under Grant No. 51479103. We are thankful to Nutonian, Inc. for enabling us to use Eureqa free of charge for academic purposes.

References

[1] Korean Concrete Institute. Thermal crack control of mass concrete, concrete practice manual (2010)

[2] Ha, J.-H.; Jung, Y.S.; Cho, Y.-g., 2014. Thermal crack control in mass concrete structure using an automated curing system. Automation in Construction. Vol.45: 16-24 .

[3] Myers T. G., Fowkes N. D. and Ballim Y., 2009. Modeling the Cooling of Concrete by Piped Water. Journal of Engineering Mechanics @ASCE .

[4] Liu C, 2004. Temperature Field of Mass Concrete in Pipe Lattice. Journal of Materials in Civil Engineering @ASCE.

[5] Bureau of Reclamation. The Study of Hoover Dam (1971).

[6] J.K. Yang, Y.Lee and J.K.Kim, 2011. Heat transfer coeffcient in flow convection of pipe cooling system in massive concrete. J.Adv.Concr.Technol. vol.9:103-114 .

[7] C. Qian and G. Gao, 2012. Reduction of interior temperature of mass concrete using suspension of phase change materials as cooling fluid. *Constr. Build. Mater.* **Vol.** *26*: 527-531 .

[8] F.Lin, X. Song, X.Gu, B. Peng and L.Yang, 2012. Cracking analysis of massive concrete walls with cracking control techniques. Constr. Build. Mater. Vol.31:12-21 .

[9] J. Yang, Y. Hu, Z. Zuo, F. Jin, and Q.Li, 2012. Thermal analysis of mass concrete embedded with double-layer staggered heterogeneous cooling water pipes. *Appl. Therm. Eng.*

[10] S.H.Chen, P. Su and I. Shahrour, 2011. Composite element algorithm for the thermal analysis of mass concrete. *Int. J. Numer. Methods Heat Fluid Flow*.**vol.** *21:* 434-447.

[11] D. Santillán, E. Salete and M.Á.Toledo, 2017. A new 1d analytical model for computing the thermal field of concrete dams due to the environmental actions. *Appl. Therm. Eng.vol.85:*160-171.

[12] J. Ding and S.Chen, 2013. simulation and feedback analysis of the temperature field in massive concrete structures containing cooling pipes. Appl. Therm. Eng. Vol. 61:554-563 .

[13] L.Zhang, C.Chang, W. Zhou, Y.Cheng and Y.Duan, 2015. Precise simulation analysis of the thermal field in mass concrete with a pipe cooling system. Appl. Therm.Eng. vol. 78:449-459 .

[14] L.Li, X.Dao and Y.Cheng, 2016. Thermal cracking analysis during pipe cooling of mass concrete using particle flow code. Adv.Mater.Sci.Eng.

[15] USACE. United States Army Corps of Engineer. Arch Dam Design. Engineering Manual, EM 1110-2-2201. Washington, DC 20314-100: 8-18 (1994).

[16] Americian Concrete Institute. ACI-207-4R-93: Cooling and Insulating system for mass concrete (1993).

[17] N.Shi, J.Ouyang and R. Zhang, 2015. Experimental study on early-age crack of mass concrete under the controlled temperature history. Adv.Mater.Sci.Eng.

[18] Americian Concrete Institute. ACI-305R-99:Hot weather Concreting (1999).

[19] Q. Li , G.Li, Y. Hu and Z. Zuo, 2014. Numerical Analysis on Temperature Rise of a Concrete Arch Dam after Sealing Based on Measured Data. Mathematical Problems in Engineering, vol. 2014 (2014) .doi: 10.1155/2014/602818

[20] B. Zhu. Thermal Stresses and Temperature Control of Mass Concrete, English reprint edition, ELSEVIER INC. and TSINGHUA UNIVERSITY PRESS: Business Park Avenue South, New York, NY 10010, USA: 362-386 (2014)

[21] I.C.Yeh, 2007. Modeling slupm flow of concrete using second-order regressions and artificial neural networks. Cem.conc.Compos. vol.29: 474-480 .

[22] V.P. Singh and Y.C., 2013. Kotiyal. Prediction of Compressive strength of Concrete using Artificial Neural Network. International Journal of Civil, Environmental, Structural, Construction and Architectural Engineering, vol.7 (12) .

[23] I.C.Yeh, 1998. Modeling the strength of high-performance concrete using artificial neural networks. Cem.conc.Compos. vol.28: 1797-1808 .

[24] J. Koza, 1992. Genetic programming, on the programming of computers by means of natural selection. MIT Press, Cambridge, MA . doi:10.1061//asce/em.1943-7889.0000046.

[25] M. Mazari and Y. Niazi, 2015. Modeling the Effect of Filler Materials on Performance of Hot Asphalt Using Genetic Programming. Airfield and Highway Pavement @ASCE .

[26] M. Schmidt and H. Lipson,2009.Distilling Free-Form Natural Laws from Experimental Data. Science, vol. 324 (5923): 81-85 .

[27] Cornell Creative Machines Lab, "Eureqa". http://creativemachines.cornell.edu/eureqa

[28] J.H. Ricketts and J.O.Carter, 2011. Estimating trend in monthly maximum and minimum temperatures in GCMs for which these data are not archive. In: 19th International Congress on Modeling and Simulation, Perth Australia, 12-16 December 2011.

[29] R. Weiand and H. Rogasik, 2015. Method for analyzing soil structure according to the size of structural elements. Computers & Geosciences. Vol. 75:96-102 .

[30] G.N. Smith, 1986. Probability and Statistics in Civil Engineering. London: Collins .

IOP Conf. Series: Materials Science and Engineering **311** (2018) 012019 doi:10.1088/1757-899X/311/1/012019

Biocalcification using Ureolytic Bacteria (UB) for strengthening Interlocking Compressed Earth Blocks (ICEB)

M M Zamer[1], J M Irwan[1], N Othman[2], S K Faisal[1], L H Anneza[1], A F Alshalif[1] and T Teddy[1]

[1]Jamilus Research Centre for Sustainable Construction (JRC), Faculty of Civil and Environmental Engineering, Universiti Tun Hussein Onn Malaysia, 86400, Parit Raja, Johor, Malaysia
[2]Micro-Pollutant Research Centre (MPRC), Faculty of Civil and Environmental Engineering, Universiti Tun Hussein Onn Malaysia, 86400, Parit Raja, Johor, Malaysia
E-mail: irwan@uthm.edu.my

Abstract. Interlocking compressed earth blocks (ICEB) are soil based blocks that allows for mortarless construction. This characteristic resulted to faster the process of building walls and required less skilled labor as the blocks are laid dry and lock into place. Recently, implementation in using bacteria as construction material improvement is vigorously used in research in order pursuit the sustainable construction works. This paper provide the results of ureolytic bacteria (UB) throughout enrichment process in soil condition to acclimatize the ICEB environment, compressive strength of 1%, 3% and 5% UB and SEM analysis of ICEB. The bacteria were added as partial replacement of limestone water in ICEB. The results showed the optimal growth achieved based on the days and absorbance from optical density (OD) test which are in 12th days with absorbance of 0.55 whereas the results for strength shows the increment of 15.25% with 5% UB on 28th days of testing compared to control specimen. Therefore this study hopes that positive results from the UB as improving in strength of ICEB which will lead to improve others ICEB properties and others construction materials.

1. Introduction

Interlocking compressed earth block (ICEB) is a cost effective, dry stacked and sustainable construction material which has potential to bring durable and affordable homes to developing counties around the world [1]. However there are also others problems associated with ICEB namely low strength, higher water absorption, low fire resistance and high porosity [2].

Previous studies had taken different approach to improve the properties of construction material such as bioconcrete [3], an addition of PET [4] and polyolefin fibers [5]. The use of bacteria is one of new fundamental research in improving construction material in order to pursuit sustainable construction. Researcher such as Muynck et al., [6] used *Bacillus sphaericus*, Dhami et al., [7] used *Bacillus megaterium*, Mukherjee [8] used *Bacillus megaterium* and Bernadi et al., [9] used *Sporosarcina paseurii*. All bacteria used by previous studies resulted in increasing on compressive strength and reduction on water absorption by comparing control sample and treated sample with bacteria. Positive results from previous studies indicated that the successfulness of using bacteria as an environmental friendly solution in improving the durability of construction material. According to Siddique & Chahal [10], bacteria are able to promote the precipitation of calcium carbonate ($CaCO_3$) in the form of calcite. These calcite acts as bio-sealant by filling the pores which lead to reduction in water absorption, porosity, permeability, enhance the strength and prevent water ingression. Hence it will improve the durability of the material properties.

Even extensive of research incorporated bacteria in construction material, less knowledge were found on adapted bacteria in ICEB. The understanding on fundamental precipitation calcium carbonate ($CaCO_3$) had been applied in this research with the use of ureolyric bactreria (UB) to improve the durability of ICEB. Therefore, this research hope that an environmental friendly solution to improve the durability and properties of ICEB would be produced by introducing the used of bacteria in ICEB.

2. Experimental works

Isolated strain UB from fresh urine was used in the biocalcification studies. The standard Nutrient Broth (NB) media were used to demonstrate and compare the efficiency of microbial induced calcite precipitation (MICP) by UB in ICEB. Enrichment process was done follows previous researcher method by Irwan et al., [11]. In order to ensure the survival of UB in ICEB environment, enrichment process was done with addition of soil sample to acclimatize the ICEB condition. The composition for optimal growth was 300ml nutrient broth added with 120ml of 40% urea, 1mg of soil substances and lastly 1 cyrogenic bead of ureolytic bacteria. Control experiment in enrichment process is flask without soil sample. Optical density of UB in both treatment and control flask were measured daily. This method to determine optimal growth condition

The UB that need to be added in the concrete will be shake in the liquid nutrient broth first until turbid following the optimum day in growth measurement. Addition of UB as partial replacement of limestone water with 1%, 3% and 5% UB are added into ICEB mix. The compressive strength (fc) was measured for 7th, 14th and 28th days. The test was carried out according to the BS 3921:1985, [12] which specify for the testing of bricks. The scanning electron microscopy (SEM) analysis was done in determination of pore size and morphology precipitation calcium carbonate ($CaCO_3$).

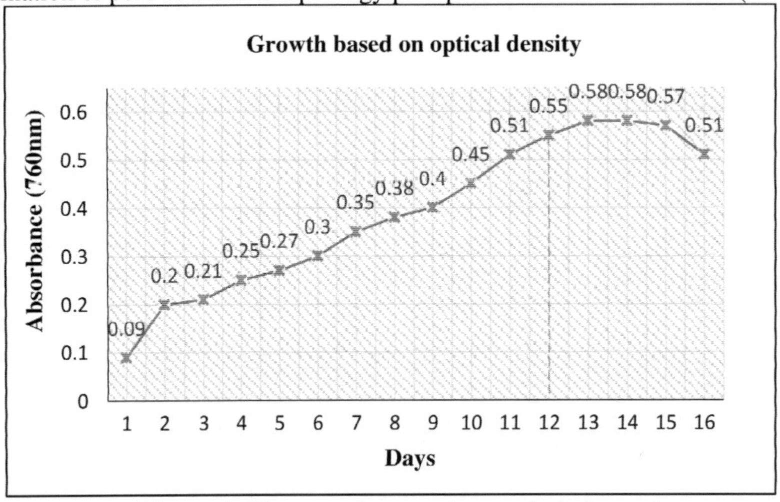

Figure 1. The growth curve for ureolytic bacteria.

3. Optimal growth profile

Survival and growth of urease producing bacteria are very essential for biocalcification activity to take place. The optimal growth of UB was monitored for 16th days in order to ensure that sufficient number of UB cell survived to allow continued biocalcification in soil condition. According to the growth chart depicted in figure 1, the highest absorbance rate was on the 13-14th day of growth. Which means, by the 12th day, UB reach optimum growth. However, when continuing to the next day which is 15th day, the amount of absorbance started to decrease. This shows that the bacteria started to become inactive and die due to depletion of nutrient provided in nutrient broth liquid [8]. Therefore to obtain the best results for bacteria growth, the number of days that was chosen are in 12th days where the results of turbidity was 0.55 absorbance. According to Siddique & Chahal [10], logarithmic phase refers to rapid growth of bacteria where the nutrient are actively being used by the bacteria. Therefore the chosen range of bacteria growth are suitable to be used in preparing the sample of ICEB.

4. Compressive strength

Figure 2 shows the results of compressive strength for 7th, 14th and 28th days of testing. The addition of UB as a liquid culture of 1%, 3% and 5% in ICEB increased the compressive strength within time compared to control specimen. The patterns of compressive strength increment are the same for all bacterial specimens. Addition of 5% UB indicates the highest increment by 15.25% in strength at 28th days compared to control sample follows with 3% of UB with 14.34% and 1% of UB with 10.71%. The highest compressive strength recorded was 6.35 N/mm^2 at the 28th days of testing for 5% addition of UB.

The positive result indicates the successfulness of UB surviving in ICEB environment and deposition of calcium carbonate, $CaCO_3$ for the strength enhancement of ICEB. The increment of compressive strength with added bacteria agreed with Dhami et al., [7] studies which state that compressive strength can be significantly increased by application of bacteria calcite.

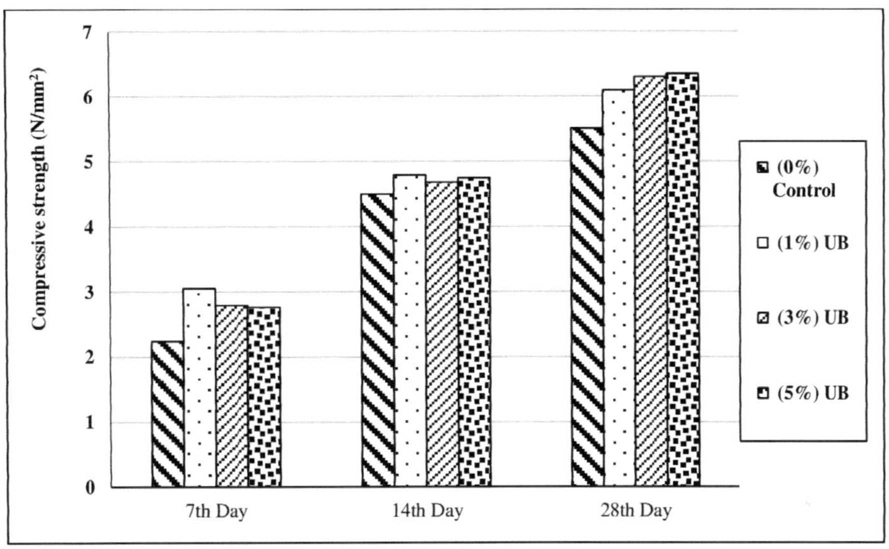

Figure 2. Compressive strength of control and bacterial ICEB.

5. SEM analysis

Microstructure analysis using SEM image shows that UB successful to grow well inside the ICEB specimens with less oxygen condition and also capable to tolerate in alkaline environment. The pore size analysis was determine by using small size of crushed ICEB passing 64 micron sieve. Figure 3 shows the average results for pore size analyses conducted on control ICEB and bacterial ICEB. Control image generally show more rough and uneven surface. This porous surface from control samples led to increase the absorption capacity of ICEB and hence decrease in durability and strength. In bacterial ICEB some part of the porous surface has been filled by calcite precipitation, thus reduction in pore size. Addition of 5% UB resulted highest reduction in average pore size with 0.643μm follows with 3% UB by 0.650μm and 1% UB with 0.772μm compared to the control specimen which was 0.839μm. There are also appearance of calcite precipitation which can be seen as smooth surface of round sphere in bacterial ICEB microstructure image. The smooth surface of round sphere indicated the bacterial forms of calcium carbonate. The average size of calcite forms was evaluated and compared between control and bacterial ICEB. As shown in figure 4, control sample which can be seen without precipitation of bacterial calcite due to 0% percentage of UB addition. Addition of 3% UB indicate highest average size of bacterial calcite form which is 12.469μm follows with 1% of UB with 9.212μm and 5% of UB by 7.815μm. The calcium carbonate precipitation was to prove that addition of UB in ICEB promote to bacterial calcite precipitation.

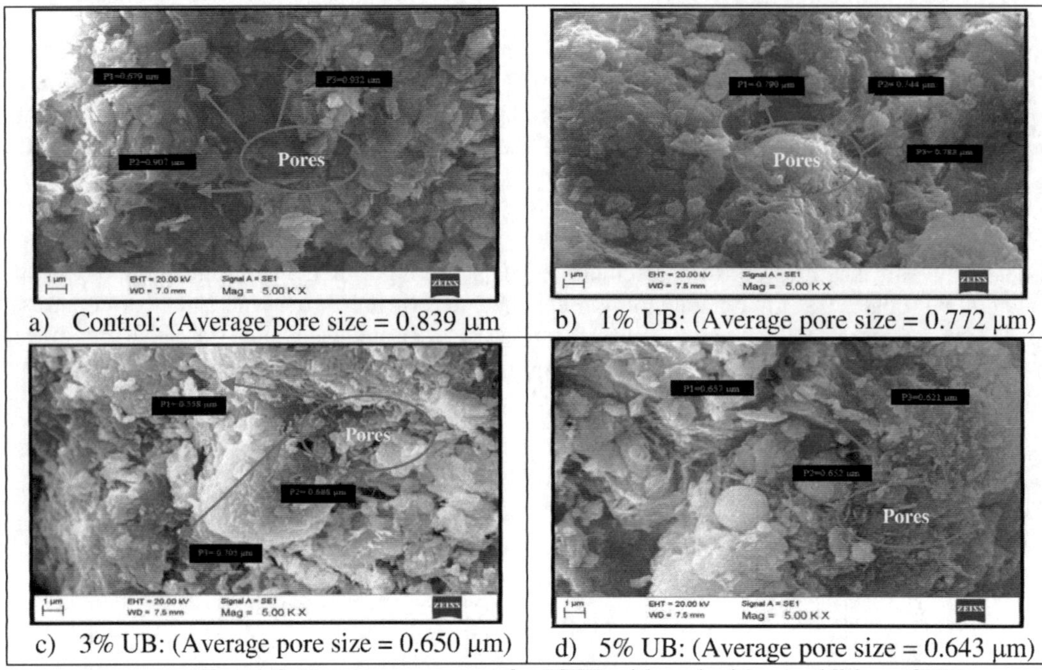

a) Control: (Average pore size = 0.839 μm

b) 1% UB: (Average pore size = 0.772 μm)

c) 3% UB: (Average pore size = 0.650 μm)

d) 5% UB: (Average pore size = 0.643 μm)

Figure 3. Pore measurement for ICEB with and without addition of UB.

a) Control: (Average size = 0)

b) 1% UB: (Average size = 9.212 μm)

c) 3% UB: (Average size = 12.469 μm)

d) 5% UB: (Average size = 7.815 μm)

Figure 4. Precipitation of calcium carbonate for ICEB with and without addition of UB.

Based on the images obtained through Scanning electron microscope (SEM), the samples which contained bacteria has shown pores being filled and surface for these samples appear more compact and smooth compared to control which is rough and porous in nature. There are also decrement in pore size of bacterial ICEB compared to control ICEB. Addition of 5% UB in ICEB indicate the highest reduction of pore size compared to the control sample by 23.36%. Therefore the results from microstructure analysis prove that by addition of UB in ICEB resulted in reducing the pores and bacterial calcite precipitation. Thus improvement the strength properties of ICEB. This findings agreed with Siddique & Chahal, [10] studies which indicated the improvement of material engineering properties due to deposition of calcium carbonate. These enchantment of ICEB strength may lead to reduction of construction cost maintenance over long period of time.

6. Conclusion
The present study suggests that addition of UB in ICEB resulted in improving the compressive strength due to biocalcification activity of UB grown in ICEB compared to control specimens. SEM analysis shows that the successful of UB in depositing the calcium carbonate and reducing the pore size of ICEB. Hence this will further improve the properties of ICEB. Positive results form addition of UB shows that the effectiveness of introducing the use of bacteria in improving the ICEB properties and potential of replacing chemical substances enchantment to achieve sustainable construction. This can favourably reduce the frequency of required maintenance for civil structures especially in bricks.

7. Acknowledgement
This research was supported by University Tun Hussein Onn Malaysia (UTHM), Malaysia, The Ministry of Higher Education Malaysia through Fundamental Research Grant Scheme (FRGS), Vot. 1577 and in collaboration with ICEB industry IRYAS INC. (M) Sdn Bhd.

8. References
[1] Laursen P T, Herskedal N A, Jansen D C and Qu B 2012 Interlocking compressed earth blocks Out-of-plane structural response, *Journal of Material and Structure*, **48**, 321-36

[2] Irwan J M, Zamer M M and Othman N 2016 A review on Interlocking Compressed Earth blocks (ICEB) with addition of bacteria, *MATEC Web Conferences*, **47**, 01017

[3] Irwan J M and Othman N 2013 An overview of bioconcrete for structural repair, *Applied mechanics and Materials*, **389**, (2013), 36-39

[4] Irwan J M, Asyraf R M, Othman N, Koh H B, Annas M M K and Faisal S K 2013 The mechanical properties of PET fiber reinforced concrete from recycled bottle wastes, *Advanced Materials Research*, **795**, 347-51

[5] Wan Ibrahim M H, Jamaludin N, Irwan J M, Ramadhansyah P J and Suraya Hani A 2014 Compressive and flexural strength of foamed concrete containing polyolefin fibers, *Advanced Materials Research*, **911**, 489-93

[6] De Muynck W, Debrouwer D, De Belie N and Verstraete W 2008 Bacterial carbonate precipitation improves the durability of cementitious materials, *Journal of Cement and Concrete Research*, **38**(7), 1005-14

[7] Dhami N K, Reddy M S and Mukherjee A 2012 Improvement in strength properties of ash bricks by bacterial calcite, *Journal of Ecological Engineering*, **39**, 31-35

[8] Mukherjee A, Dhami N K, Reddy B V V and Reddey M S 2013 Bacterial Calcification for Enhancing Performance of Low Embodied Energy Soil-Cement Bricks, *Journal of Sustainable Construction Material and Technology*, **1**, 1-10

[9] Bernadi D, Dejong J T, Montoya B M and Mattinez B C 2014 Bio-bricks: Biologically cemented sandstone bricks, *Journal of Construction and Building Material*, **55**, 462-69

[10] Siddique R and Chahal N K 2011 Effect of ureolytic bacteria on concrete properties, *Journal of Construction and Building Materials*, **25**(10), 3791-801

[11] Irwan J M, Anneza L H, Othman N, Husnul T and Alshalif A F 2016 Isolation and Identification of Concrete Environment bacteria, *IOP Conference Series: Materials Science and Engineering*, **136**(1), 012088

[12] British Standard Institution, BS 3921: 1985, Specification for Clay Bricks, Clay Product Standards Committee, London, England

IOP Conf. Series: Materials Science and Engineering **311** (2018) 012020 doi:10.1088/1757-899X/311/1/012020

Mechanical Design of Metal Dome for Industrial Application

Thomas Jin-Chee Liu, Li-Wei Chen and Nai-Pin Lin

Department of Mechanical Engineering, Ming Chi University of Technology, Taishan, New Taipei City, Taiwan
E-mail: jinchee@mail.mcut.edu.tw

Abstract. In this paper, the mechanical design of metal domes is studied using finite element analysis. The snap-through behavior of a practical button design that uses a metal dome is found. In addition, the individual click ratio and maximum force for a variety of metal domes are determined. This paper provides guidance on button design for industrial engineers.

1. Introduction

The metal dome shown in Fig. 1 is a type of thin shell structure, which is known as a spherical cap in solid mechanics. This element is usually applied in the button design of electronic products such as cameras, mobile phones, or audio players. Due to the snap-through/buckling deformation typical of a metal dome, one can feel the jumping or clicking motion when pressing the button.

Using either an analytical or a numerical method, the snap-through deformation of a spherical cap or arch shell has been extensively studied [1-7], while their applications were discussed by Tomitsuka et al. [5]. However, the practical application and discussion of a metal dome for button design was presented in very few references.

In this paper, non-linear finite element analysis will be applied for a metal dome used in a button design. The snap-through deformation and click ratio will be obtained for some practical design cases. Design guidelines will be proposed on the basis of the results of this research.

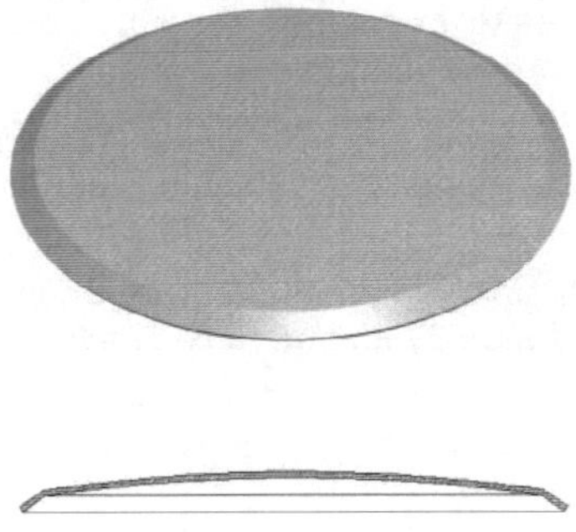

Figure 1. Metal dome.

2. Problem definition and finite element model

The geometry of the metal dome to be analyzed is shown in Fig. 2. It is composed of a curved cap and flat ring. The dome is a thin-shell structure and made of high-strength stainless steel (JIS SUS 301 EH). Its main dimensions are R, Q, H, and t. The indenter with radius r experiences a downward force and displacement that cause it to press into the metal dome. In Table 1, the material constants of JIS SUS 301 EH are listed, while the elasto-plastic stress–strain curve is shown in Fig. 3. Due to its relative rigidity, the indenter is assumed to be a rigid body. The metal dome is constrained on *L2* and subjected to the load from the indenter.

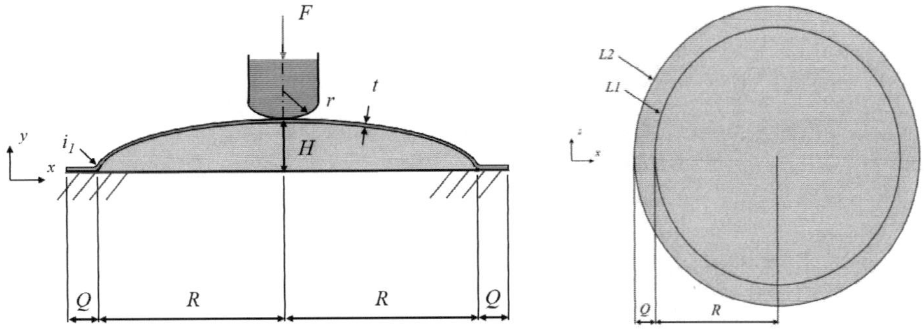

Figure 2. Geometry of metal dome.

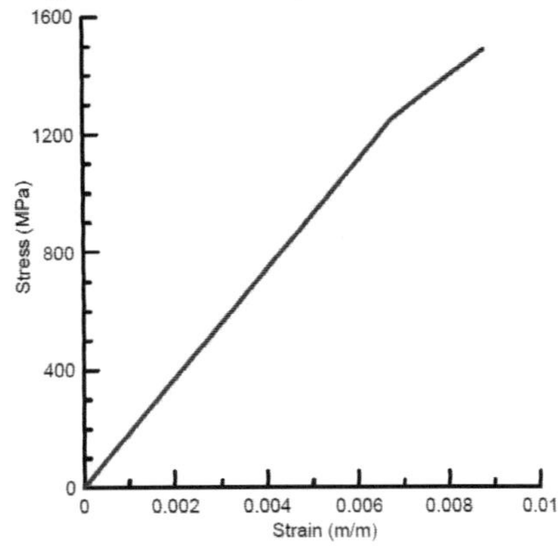

Figure 3. Elasto-plastic stress-strain curve of stainless steel (JIS SUS 301 EH).

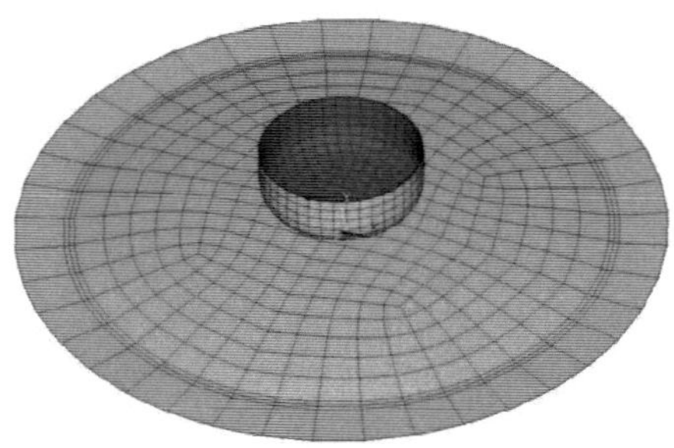

Figure 4. Finite element model.

Table 1. Material properties of stainless steel (JIS SUS 301 EH).

Material	Young's modulus E (GPa)	Poisson's ratio ν	Yielding strength S_y (MPa)
JIS SUS 301 EH Stainless steel	186	0.30	1250

ANSYS software is employed for the non-linear finite element analysis. The contact condition between the dome and indenter is considered as well as the large displacement and elasto-plastic stress–strain curve. For the finite element simulation in ANSYS, the element types used are as follows: SHELL93, TARGE170, and CONTA174.

A typical finite element model is shown in Fig. 4. The indenter has a prescribed displacement along the *y*-direction, which can be downward or upward to control its motion.

3. Design rules

The snap-through behavior (path *a-b-c*) of the metal dome is shown in Fig. 5. In this figure, *F* and *v* represent the applied force and indentation displacement, respectively. Along the snap-through path *a-b-c*, one can feel a click or jump motion when the dome or button is pressed.

The click ratio R_c is defined as follows:

$$R_c = (F_1 - F_2) / F_1 * 100\% \quad (1)$$

R_c is an important parameter for the metal dome or button. A larger R_c value represents a more obvious jump motion when the button or dome is pressed. Referring to past studies [5], the click ratios are designed within a range of 30% to 60% to achieve a satisfying feeling for the user when the button is pressed. Past studies [5,8] also proposed that the range of maximum force F_1 should be designed to be from 1 N to 5 N for mobile communication products.

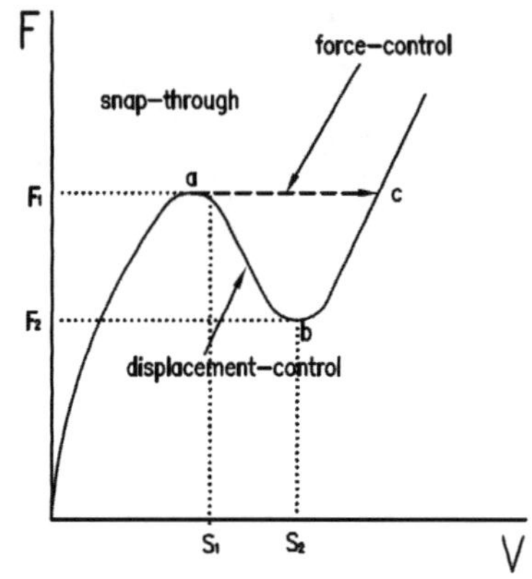

Figure 5. Snap-through behavior.

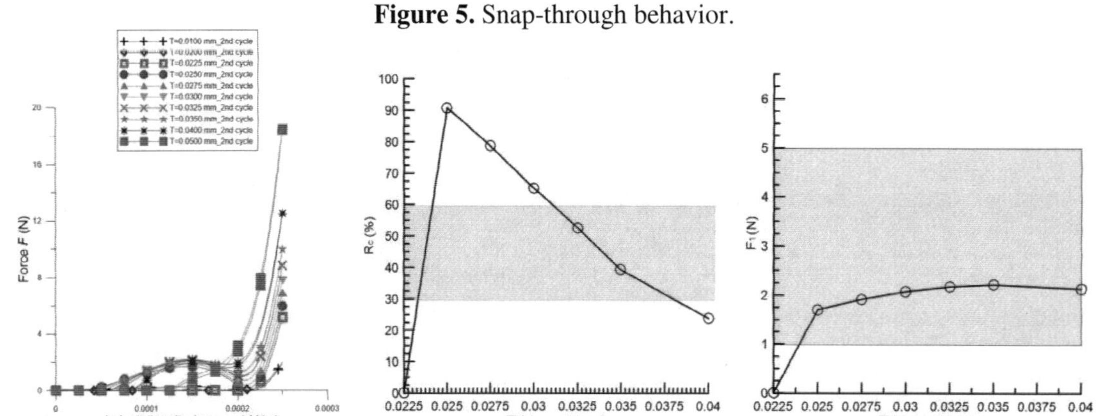

(a) (b) (c)

Figure 6. (a) *F-v* relation, (b) R_c variation, (c) F_1 variation. (effects of *t*)

4. Numerical results

First, the effects of the thickness on the click ratio (R_c) and maximum force (F_1) are discussed. The dimensions R = 1 mm, Q = 0.2 mm, H = 0.1 mm, i_1 = 0.2 mm, and r = 0.3 mm are kept constant. In Fig. 6, the *F–v* relation, R_c variation, and F_1 variation for different thicknesses are obtained from the finite element analysis.

The color bands in figures (b) and (c) denote the design ranges according to past research [5,8]. Only the cases of t = 0.0325 mm and t = 0.035 mm meet the design requirements. For the cases of $t \leq$

0.0225 mm, the snap-through behavior is lost and the click ratio cannot be defined. In summary, the click ratio decreases when the thickness increases.

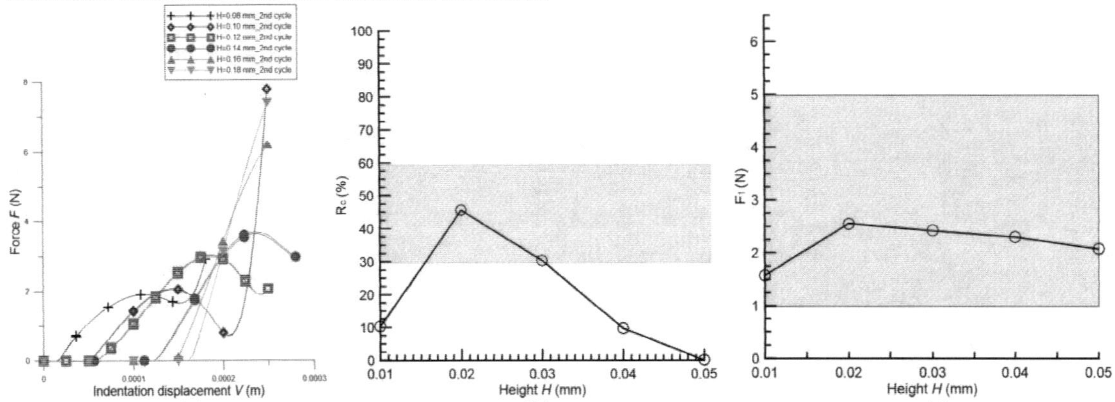

(a) (b) (c)

Figure 7. (a) F-v relation, (b) R_c variation, (c) F_1 variation. (effects of H)

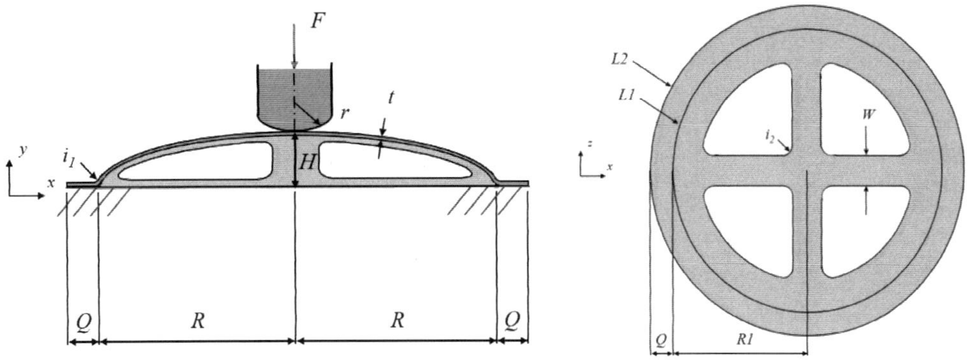

Figure 8. Special case.

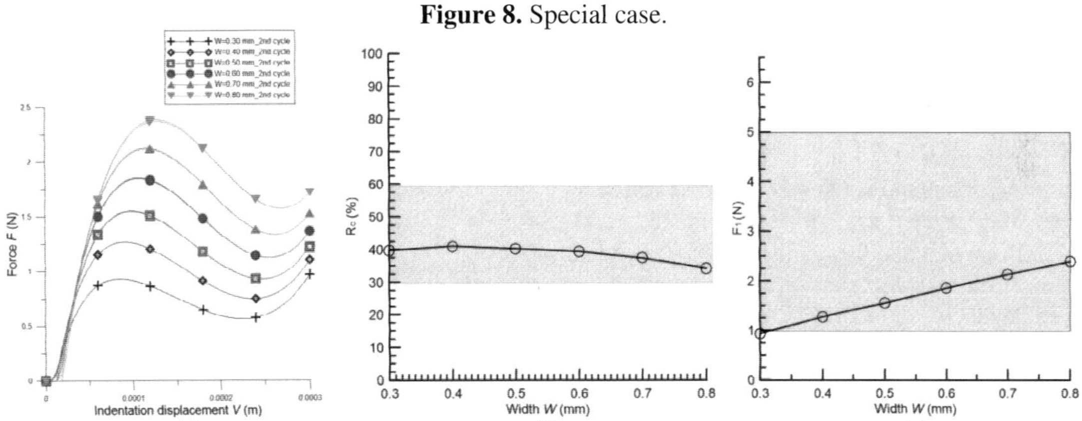

(a) (b) (c)

Figure 9. (a) F-v relation, (b) R_c variation, (c) F_1 variation. (effects of W)

The effects of H are also discussed. The dimensions R = 1 mm, Q = 0.2 mm, t = 0.03 mm, i_1 = 0.2 mm, and r = 0.3 mm are kept constant; the results are shown in Fig. 7. Only the cases of H = 0.02 mm and H = 0.03 mm meet the design requirements. In figures (b) and (c), we see that the case of H = 0.02 mm gives the peak value.

A special case for a metal dome with four holes is shown in Fig. 8. The dimensions R = 1.95 mm, Q = 0.3 mm, H = 0.2 mm, t = 0.05 mm, i_1 = 0.2 mm, and r = 0.3 mm are kept constant. The dimension W is a variable.

The results of these cases are shown in Fig. 9; they nearly meet the design requirements. Similar *F–v* curves are plotted as for the previous case. We can see that the click ratios vary slightly with different *W* values. However, F_1 consistently increases as the width increases.
In addition, Fig. 10 shows the stress and deformation of the metal dome with $W = 0.7$ mm.

(a) (b)

Figure 10. Contour of von Mises stress under (a) maximum indentation and (b) unloading condition. (units of stress: Pa)

5. Conclusion

The finite element results provide information and a methodology for metal dome and/or button design. The ideal snap-through behavior, click ratio, and maximum force of various metal domes were obtained. The mechanical behavior of the dome was shown to be affected by the geometric condition. By controlling the shape, the metal dome can meet the design requirements. In particular, the disappearance of the snap-through behavior for some domes must be avoided. In summary, the JIS SUS 301 EH stainless steel is a good choice for metal dome design.

6. References

[1] Brush D O and Almroth B O 1975 *Buckling of Bars, Plates, and Shells* (New York: McGraw-Hill)
[2] Wempner G and Talaslidis D 2003 *Mechanics of Solids and Shells: Theories and Approximations* (New York: CRC Press)
[3] Mescall J F 1965 *ASME J. Appl. Mech.* **32** 936
[4] Bathe K J and Ozdemir H 1976 *Comput. Struct.* **6** 81
[5] Tomitsuka T, Ajimura S, Kawahira T and Noguchi Y 2003 *Fujikura Tech. Rev.* **104** 42 (in Japanese)
[6] Sankar A, El-Borgi S, Ben Zineb T and Ganapathi M 2016 *Composites Part B: Engineering* **99** 472
[7] Soltanieh G, Kabir M Z and Shariyat M 2017 *Composite Structures* **180** 581
[8] Ma R H 2011 *Human Factor Engineering* (Beijing: Peking University Press) (in Chinese)

IOP Conf. Series: Materials Science and Engineering **311** (2018) 012021 doi:10.1088/1757-899X/311/1/012021

Biochar Preparation from Simulated Municipal Solid Waste Employing Low Temperature Carbonization Process

C Areeprasert, P Leelachaikul, G Jangkobpattana, K Phumprasop and T Kiattiwat

Department of Mechanical Engineering, Faculty of Engineering, Kasetsart University, 50 Ngam Wong Wan Road, Lat Yao, Chatuchak, Bangkok 10900, Thailand.
E-mail: fengcta@ku.ac.th

Abstract. This paper presents an investigation on carbonization process of simulated municipal solid waste (MSW). Simulated MSW consists of a representative of food residue (68%), plastic waste (20%), paper (8%), and textile (4%). Laboratory-scale carbonization was performed in this study using a vertical-type pyrolyzer varying carbonization temperature (300, 350, 400, and 450 °C) and heating rate (5, 10, 15, and 20 °C/min). Appearance of the biochar product was in black and the volume was significantly reduced. Low carbonization temperature (300 °C) might not completely decompose plastic materials in MSW. Results showed that the carbonization at the temperature of 400 °C with the heating rate of 5 °C/min was the optimal condition. The yield of biochar from the optimal process was 50.6% with the heating value of 26.85 MJ/kg. Energy input of the process was attributed to water evaporation and the decomposition of plastics and paper. Energy output of the process was highest at the optimal condition. Energy output and input ratio was around 1.3-1.7 showing the feasibility of the carbonization process in all heating rate condition.

1. Introduction

One of the consequences of growing population and economic expansion is a waste management problem. Urbanization causes an increase of municipal solid waste (MSW). A systematic management of MSW is necessary to maintain good condition of the city or village. MSW is produced by households or residential areas, community, commercial area, markets, and other human activities. It excludes industrial, hazardous, construction, and demolition waste. In general, the MSW composition is organic that is biodegradable material such as food residues and yard waste; paper waste including scraps paper, newspapers, magazines, bags, packaging paper; plastic waste such as bottle, packaging, containers, and bags; glass product, for example, bottles and containers; metal in form of cans, foils, and other containers or parts; others could be textiles, leathers, ash, rubber, other inert materials.

Considering waste-to-energy routes (WTERs) as a solution, waste materials turn out to be attractive assets. Unusable waste materials, such as municipal solid waste from residence [1,2] can be used for energy purpose. Alternative to waste incineration that is one of the most common method of waste elimination, carbonization process is a promising conversion process. It does not convert the MSW directly to energy but turn them into a valuable material that can be handle easily. After carbonization, the biochar product can be used for many applications such as solid biofuel and soil amendment. Biochar is generally produced from agricultural waste materials or biomass using carbonization process. Biomass residues, apricot stone, hazelnut shell, grapeseed, and chestnut shell, have been used to produce biochar and bio-oil and it was found that the biochar products contained high carbon content and high heating value while the bio-oil needs some upgrading process before utilization [3] This paper investigated biochar production from simulated MSW using carbonization process. Major composition of MSW sample was tested using representative materials. Carbonization process was performed at laboratory-scale level. Temperature and heating rate of the carbonization were varied. Appearance and product yield were studied to find the optimal condition. Moreover, energy input, energy output, and energy output to input ratio were calculated. This process analysis can give information on feasibility of the process.

2. Material and methods

2.1. Raw materials

Sample used in this study was a simulated municipal solid waste (MSW) as shown in Fig. 1. The major waste composition in Thailand based on our previous study was organic, plastic, paper, and textile . Thus, the simulated MSW composition was food residue (68%), plastic waste (20%), paper (8%), and textile (4%) [4]. Plastic sample consists of PET 4%, HDPE 12%, and PP 4%. Chemical composition of the raw sample from elemental analysis (ASTM D5373) was 43.4% carbon, 6.3% hydrogen, 1.4% nitrogen, and 48.8% oxygen and from proximate analysis (ASTM D7582) was 77.8% of volatile matter, 14.6% of fixed carbon, and 7.6% of ash. Moisture of the raw sample was 65% and heating value (ASTM D5865) was around 20 MJ/kg in dry basis. Raw material was cut into small size roughly 1-2 cm while its density was very low.

Figure 1. Simulated municipal solid waste for carbonization process.

2.2. Carbonization

Carbonization experiment was done at laboratory-scale level using a vertical-type pyrolyzer with the PID controller and other equipment as shown in Fig. 2. For each experiment, the sample was filled in the reactor about 50 g and the nitrogen gas was injected 200 ml/min. The target temperature was (300, 350, 400, and 450 °C) and heating rate (5, 10, 15, and 20 °C/min). The duration of carbonization process was about 1 hour. The carbonization test was firstly performed at the heating rate of 25 °C/min with variation of temperature to check the optimal temperature condition then the variation of the heating rate was done for final optimization. After finished the experiment, the reactor was left to cool down. Then, biochar was taken out from the reactor. Each condition was duplicated.

2.3. Energy balance

Energy input of the process was energy used to heat the sample including sensible heat of solid phases (Eq.1), sensible heat of liquid phase (Eq.2), latent heat of melting (Eq.3) and latent heat of vaporizing (Eq.4) of each material. Summation of Eq.1, Eq.2, Eq.3, and Eq.4 obtains total energy need for carbonization process. The assumption of the parameters in all equations is presented in Table 1 where

Q is an amount of heat (kJ); m is mass (kg); T_m is melting temperature (°C); T_g is vaporization temperature (°C), T_0 is room temperature (°C); L_m is latent heat of melting (kJ/kg); and L_g is latent heat of vaporization (kJ/kg); c_p is specific heat capacity (kJ/kg•K).

$$Q_s \text{ (solid)} = mc_p(T_m-T_0) \quad (1)$$

$$Q_s \text{ (liquid)} = mc_p(T_g-T_m) \quad (2)$$

$$Q_m = mL_m \quad (3)$$

$$Q_g = mL_g \quad (4)$$

Energy output from the process was energy in biochar and calculated by multiplying mass of biochar product and heating value of the sample. Energy output to input ratio was finally calculated. It determined feasibility of the process.

Figure 2. Laboratory-scale apparatus for carbonization process: (1) Pyrolyzer; (2) Nitrogen gas tank; (3) Piping system; (4) Exhaust hood; (5) PID controller; (6) Gas condenser.

Table 1. Summary of parameters and energy input [5-7].

	$c_{p,\ solid}$ [kJ/kg·K]	$c_{p,\ liquid}$ [kJ/kg·K]	L_m [kJ/kg]	L_g [kJ/kg]	T_m [°C]	T_g [°C]	Q_s (solid) [kJ]	Q_s (liquid) [kJ]	Q_m [kJ]	Q_g [kJ]	Q_{total} [kJ]
PP	1.6	2.1	209	1900	160	354	0.4	0.8	0.4	3.8	5.5
PET	1.1	1.6	139	1400	260	392	0.5	0.4	0.3	2.8	4.0
HDPE	2	2.7	245	2200	130	380	1.3	4.1	1.5	13.2	20.0
Textile	1.3	-	-	3200	-	300	0.7	0	0	6.4	7.1
Paper	1.2	-	-	3200	-	300	1.3	0	0	12.8	14.1
Food	1.4	-	-	3200	-	300	0.8	0	0	6.3	7.1
Water	-	4.2	-	2300	-	100	0	10.08	0	73.6	83.7

3. Results and Discussion

3.1 Appearance of the product

Appearance of biochar Fig. 3 illustrates the appearance of biochar obtained from carbonization process. Biochar product was relatively uniform and its color was black. When comparing the effect of carbonization temperature, the overall appearance was similar; however, low temperature carbonization (300 °C) cannot decompose the plastic material as the plastic sample was melt and accumulated (data not shown). The biochar obtained from simulated MSW was easier for grinding compared to the original material. For example, the plastic HDPE or PP from plastic bag or plastic container are difficult to reduce the size. Thus, using carbonization process can improve physical property of the material. The volume of the final product was significantly reduced.

Figure 3. Appearance of biochar from carbonization process of simulated MSW

3.2 Yield of the product

Effect of temperature and heating rate on yield of the biochar at the beginning, the heating rate was fixed at 25 °C/min and the carbonization temperature was varied. Yield of the biochar is presented in Fig. 4a. From production curve, the carbonization temperature that gave the highest biochar yield was 400 °C at 20.9% (dry basis yield). When the carbonization temperature was further increased to 450 °C, the yield of biochar was reduced. As temperature increases, the volatile matter in the sample released faster resulting in the increasing of gaseous products. These gaseous products can be either condensable or non-condensable gas and they could be recovered for additional value.

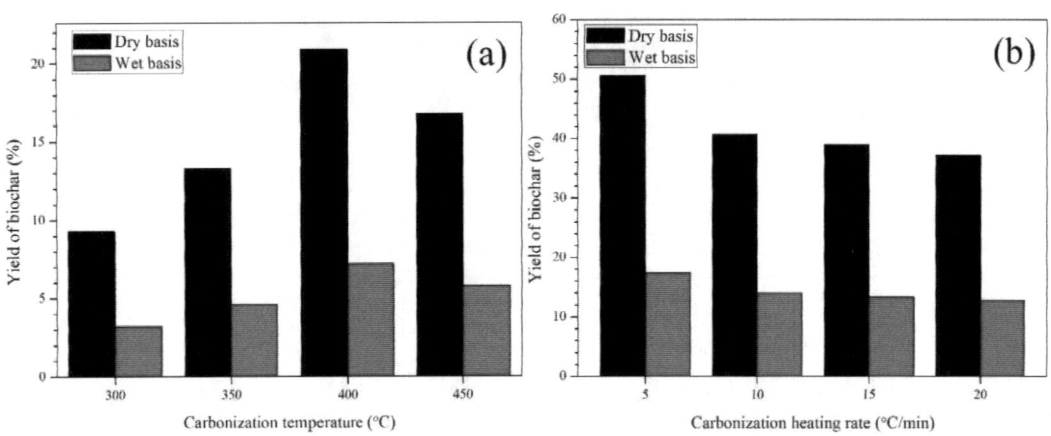

Fig. 4 Yield of biochar from carbonization process: (a) at 25 °C/min with variation of temperature; (b) 400 °C with variation of heating rate.

Then, the heating rate of the carbonization process was varied at the temperature of 400 °C and the results show in Fig. 4b. Yield of biochar can be increased to 50.6% at the heating rate of 5 °C/min while the increasing of heating rate reduces the yield of the product. In general, the high heating rate was used for pyrolysis process aiming to obtain the oil product. Thus, to produce biochar, low heating rate was preferred.

3.3 Energy balance of the process

Detail calculated results of energy input is presented in Table 1. It can be observed that the largest amount of energy was utilized for water evaporation while heating HDPE and paper consumed significant amount of energy. Energy balance of the process is shown in Table 1. Recovered energy in biochar was around 26-29 kJ/kg and the higher heating rate could slightly increase the heating value of the biochar. However, high heating rate negatively affected the yield of biochar. Since lower yield of biochar was obtained when increases the heating rate, the energy yield in the product was lowered. It should be noted that the energy output from the process was not only in the biochar, but also in the gas or recovered oil. These products could increase the energy output of the system. The process with 5 °C/min heating rate was the optimal as the energy output was the highest. This resulted in the highest energy output/input ratio of 1.65 showing highest feasibility of the process. High heating rate condition gave net energy output from the carbonization process.

Table 2. Energy balance of the process

Heating rate [°C/min]	Heating value of biochar [MJ/kg]	Energy output [MJ/kg MSW]	Energy input [MJ/kg MSW]	Energy output/input ratio
5	26.85	4.67	2.83	1.65
10	28.45	3.98		1.41
15	29.42	3.94		1.39
20	28.38	3.63		1.28

4. Conclusion

Carbonization process was applied to simulated MSW that consists of food residue (68%), plastic waste (20%), paper (8%), and textile (4%). Carbonization was performed at the temperature of 300, 350, 400, and 450 °C and the heating rate of 5, 10, 15, and 20 °C/min. Energy balance of the process was investigated to observe the feasibility of the process. Conclusion can be made as follows:

1. Appearance of the biochar obtained from carbonization process was in black and the volume was significantly reduced. Low carbonization temperature might not completely decompose plastic materials in MSW.

2. Optimal temperature in this study was 400 °C with the heating rate of 5 °C/min. This condition gave the highest yield of biochar at 50.6% dry basis. The heating value of biochar from this optimal condition was 26.85 MJ/kg.

3. Increase the carbonization temperature could help increase the yield of the biochar; however, too high temperature gave the opposite result. Lower heating rate can increase the biochar yield.

4. Energy input of the process was attributed to water evaporation and the decomposition of plastics and paper. Energy output of the process was highest at the optimal condition. Energy output and input ratio was around 1.3-1.7 showing the feasibility of the carbonization process in every heating rate condition.

5. References

[1] W. Tsai, Y. Chou: Renew. Sust. Energ. Rev. Vol. 10 (2006), p. 491.

[2] H. Cheng, Y. Hu: Bioresour. Technol. Vol. 101 (2010), p. 3816.

[3] D. Özçimen, A. Ersoy-Meriçboyu. Renew. Energy Vol. 35.6 (2010), 1319.

[4] C. Areeprasert, J. Kaharn, B. Inseemeesak, P. Phasee, C. Khaobang, A. Kuhavichanun, P. Theerarojprateep, W. Siwakosit: J. Mater. Cycles Waste (2017), inpress.

[5] Information on http://polymerdatabase.com

[6] Information on http:// www.tainstruments.com/pdf/literature/TN048.pdf

[7] Information on https://www.fire.tc.faa.gov/pdf/05-14.pdf

IOP Conf. Series: Materials Science and Engineering **311** (2018) 012022 doi:10.1088/1757-899X/311/1/012022

A Pontential Agriculture Waste Material as Coagulant Aid: Cassava Peel

N. Othman, N-S Abd- Rahim, S-N-F Tuan-Besar, S. Mohd-Asharuddin, and Kumar, V

Micro-Pollutant Research Centre (MPRC), Faculty of Civil and Environmental Engineering, Universiti Tun Hussein Onn Malaysia, 86400, Parit Raja, Johor, Malaysia
E-mail: norzila@uthm.edu.my

Abstract. All A large amount of cassava peel waste is generated annually by small and medium scale industries. This has led to a new policy of complete utilization of raw materials so that there will be little or no residue left that could pose pollution problems. Conversion of these by-products into a material that poses an ability to remove toxic pollutant would increase the market value and ultimately benefits the producers. This study investigated the characteristics of cassava peel as a coagulant aid material and optimization process using the cassava peel was explored through coagulation and flocculation. This research had highlighted that the Cassava peels contain sugars in the form of polysaccharides such as starch and holocellulose. The FTIR results revealed that amino acids containing abundant of carboxyl, hydroxyl and amino groups which has significant capabilities in removing pollutants. Whereas analysis by XRF spectrometry indicated that the CP samples contain Fe_2O_3 and Al_2O_3 which might contribute to its coagulation ability. The optimum condition allowed Cassava peel and alum removed high turbidity up to 90. This natural coagulant from cassava peel is found to be an alternative coagulant aid to reduce the usage of chemical coagulants

1. Introduction

Principle of water treatment brings raw water up to drinking water quality. Water treatment process for surface water supplies include the basic water treatment processes namely, screening, mixing, coagulation & flocculaation, sedimentation, filteration and chlorination. Coagulant used to be applied in coagulation & flocculation process to remove priority pollutant that is solid. Alum is the most favorable coagulant used in various water treatment plant in Malaysia due to its low cost [1].
Despite the superiority of chemical coagulants in treating turbid water, knowledge is still lacking in terms of green chemistry. Several studies had published the detrimental effects of chemical coagulants on human health [2]. Epedemiological studies and clinical observations have suggested at least 70% of positive correlations between the presence of aluminium in drinking water and Alzheimer's disease [2]. Therefore, there is a high demand to find an alternative coagulant which is preferably from natural, renewable sources and safe for human health. Several studies have been done on natural coagulants produced or extracted from plants, animals, or microorganisms [3][4]. Inorganic coagulants such as alum in combination with lime have been conventionally used for removal of turbidity from surface waters. The sludge formed from such treatment poses disposal problems because of its aluminium content and tend to accumulate in the environment and also because of large volume. Therefore, it is desirable that other cost effective and more environmentally acceptable alternative coagulants be developed to supplement if not replace alum, ferric salts and synthetic polymers. In this context, natural coagulants present viable alternatives for developing countries.

Natural macromolecular coagulants are promising and have attracted the attention of many researchers because of their abundant source, low price, multi-purposeness and biodegradation. Okra, Rice, Moringa olifera and chitosan are natural compounds which have been used in turbidity removal [5]. In this study Cassava peel is among non-exploited waste to be selected as a coagulant aid. Cassava with high protein and carbohydrate is among the criteria to be selected as coagulant aid.

2. Experimental Work

2.1 Raw Water
Samples of raw water were collected from raw water inlet at Sembrong water treatment plant, Sembrong, Johor, Malaysia.

2.2 Preparation of Cassava Pee l Starch (CPS) as coagulant aid
The cassava peel (CP) used in this study was obtained from small medium industries around Parit Raja, Johor, Malaysia. The peels were cleaned using tap water before rinse with distilled water. After that, periderm layer was scrapped off from the whole peel using a fruit scraper. The peel then chopped into approximately 1 cm length and pulverized. Subsequently, the pulp was suspended in 10 times of its volume of distilled water and stirred for 2 minutes. The suspension was then filtered using doubled fold muslin cloth. The filtrate was allowed to stand for 2 hours for sedimentation before the top liquid was decanted and discarded. The sediment was collected and oven dried at 60°C for 24 hours to remove the moisture content. Dried CPS was kept in air tight plastic container for further use [2].

2.3 Characterization of Cassava Peel
SEM-EDX- This analysis was conducted to determine the surface morphology of the CP samples. A scanning electron microscope (SEM) equipped with energy dispersive X-ray spectrometer (EDX).
Fourier Transform Infrared (FTIR) spectroscopy- was done to identify surface functional group of native CP samples and performed using Perkin Elmer Spectrum 100 [2]. The adsorption bands were recorded at characteristics wave numbers between 600cm^{-1} and 4,000 cm^{-1}[6]
XRF- Native cassava peel were subjected to X-Ray Fluoresence analysis to determine elemental composition of the samples using Bruker AXS S4 Pioneer XRF spectrometer [7]

2.4 Coagulation and Flocculation Test
Three set of JAR test were carried out. Optimization was investigated in the effect of coagulant dosage: alum (5, 10, 15, 20, 25, 30 mg/L), CP (100, 200, 300, 400, 500, 600mg/L), alum + CP(100%:0%, 70%:30%, 50%:50%,30%:70%,0%:100%). The pH and settling time were kept constant at 7 and 60min.

2.5 Preparation of CPS and Alum Stock Solution
Stock solution of 1000mg/L for CPS was prepared by adding 1g of CPS powder into 1L of distilled water. The suspension was stirred for 1 hour in order to ensure the powder mix uniform of the particles.
Alum solution was prepared by dissolving 10 g Aluminium Sulphate ($Al(SO_4)_3.18H_2O$) in 1L of distilled water. This stock solution was used to prepare alum solution of varying concentration for coagulation test.

3. Results and Discussions

3.1 (a) Characterization of Cassava Peel
Figure 1 shows that the Cassava Peel surface feature is non-porous and heterogeneous characteristics. Similar observations were reported by other studies on SEM analysis on other fruit waste [8]. Smooth and globular in shape of bound starch granules were observed to mostly cover overall surface of the cassava peel.

Figure. 1. FE-SEM micrograph of native Cassava Peel

Various chemical groups and bonds in accordance with the respective wavenumbers (cm^{-1}) were shown as peaks in the FTIR spectra In Figure 2, the broad band observed between 3500 – 3200 cm^{-1} was assigned to O-H groups of free hydroxyl groups and bonded O-H group in polymeric compounds such as alcohols, phenols and carboxylic acids presented in pectin, cellulose and lignin on the CP. The peak at wavenumber between 3000 – 2850 cm^{-1} may indicate the stretch of symmetric or asymmetric CH, CH_2, of aliphatic acids [9]

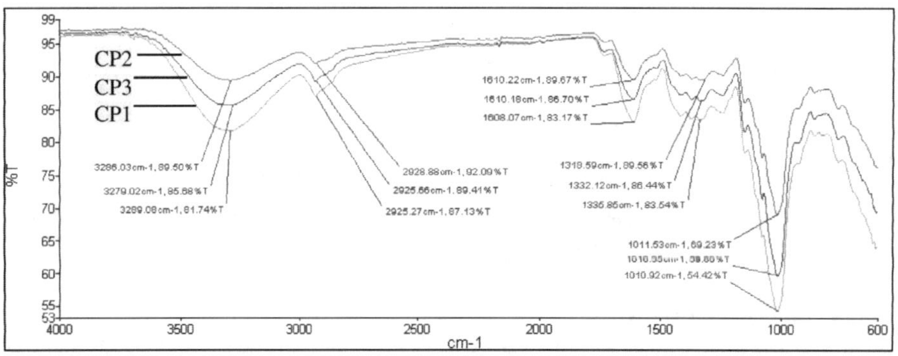

Figure. 2. FTIR spectra of CP1, CP2 and CP3 of native Cassava Peel

Furthermore, the characteristic peak appeared between 1750 – 1680 cm^{-1} are demonstrating the stretching vibration of C=O bond of carboxyl groups. Whereas the peaks observed between `1375 – 1300 cm^{-1} may reflect the stretching vibration of ionic carboxylic groups of pectin. While the deep peaks between 1300 – 1000 cm^{-1} indicate the C-O stretching of –COOH [8].

XRF analysis results are presented in Table 1. The elements present in the CP that gave it the coagulant properties are Al_2O_3 and Fe_2O_3. Both elements are expected to give the CP the ability to aid the coagulation process via precipitation of the particles in the raw water.

Table 1. Elemental composition of native cassava peel

Formula	Weight (%)
	CPS
C	0.10
O	5.83
CaO	3.48
Fe_2O_3	0.41
SiO_2	0.40
SO_3	0.94

IOP Conf. Series: Materials Science and Engineering **311** (2018) 012022 doi:10.1088/1757-899X/311/1/012022

Al_2O_3	0.26
P_2O_5	0.42
LOI	88.16

3.1(b) Characterization of floc

Figure 3 shows that the alum floc in the form of shaped pieces with veined and no regular shape was detected. The SEM image of floc is in a granular shape (smooth and globular in shape of bound starch granules) by using cassava peel starch (CPS) (Figure 4). Figure 5 shows pieces of floc for alum + CPS. There are also formations of shining particle that indicate on metals adsorbed onto the floc [10].

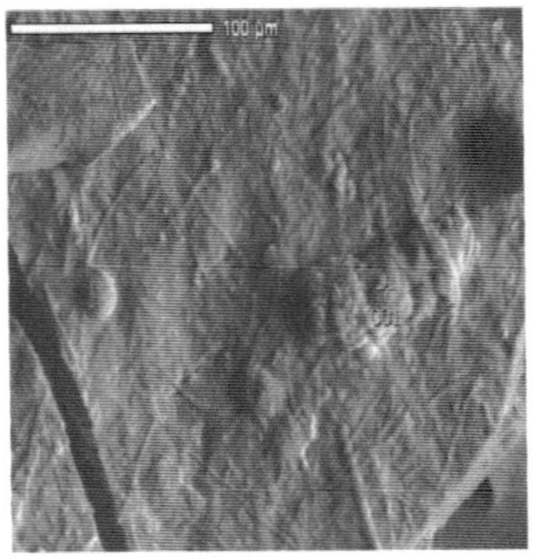

Fig. 3. Alum Floc

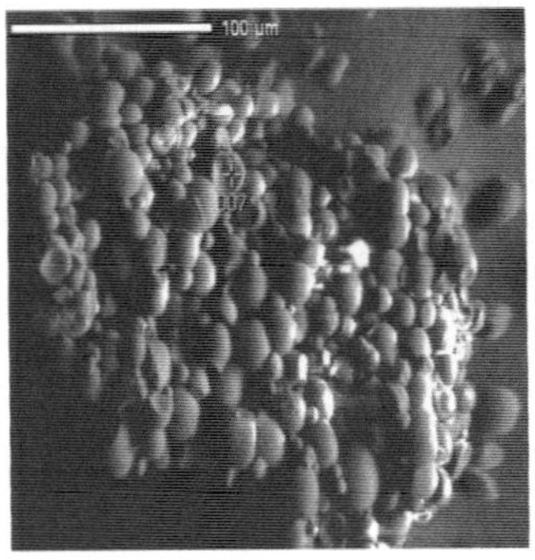

Fig. 4. CPS floc

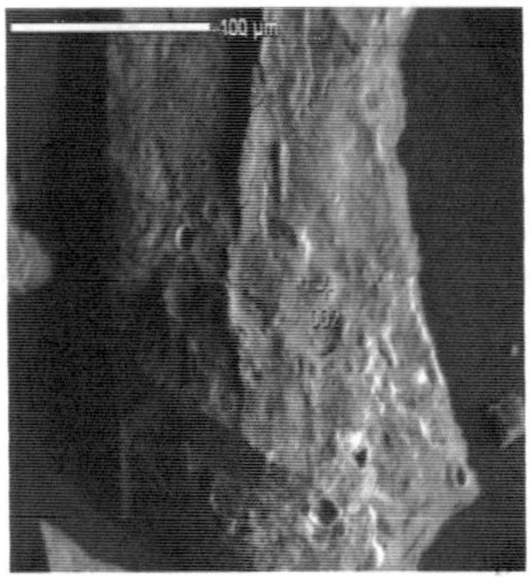

Fig. 5. Alum + CPS floc

(Figure 3,4 and 5: Floc after coagulation flocculation process)

3.2Effect of Dosage

Figure 6 shows turbidity removal increase up to 86.04% with increasing of alum dosage. The percentage of chemical oxygen demand (COD) removal was increased from 23.08% to 52.31% with the addition of alum dosage. Alum is high in charge density and it is dissolved in the suspension [6] with the mechanism on charge neutralization [10].

Figure 7 shows the effect of CP dosage. The percentage of turbidity removal was decreasing from 39.61% up to the lowest 23.64% with additional CP dosage. Natural coagulant is used as a coagulant aid to bridge the coagulated particles when aluminium or iron salt has been used as a primary coagulant in water and wastewater treatment [3].

Figure 8 shows the effect of aluminium sulphate + cassava peel starch dosage. The turbidity reduce from 78.05% to 32.20% at ratio 30 : 70 to 0 : 100.

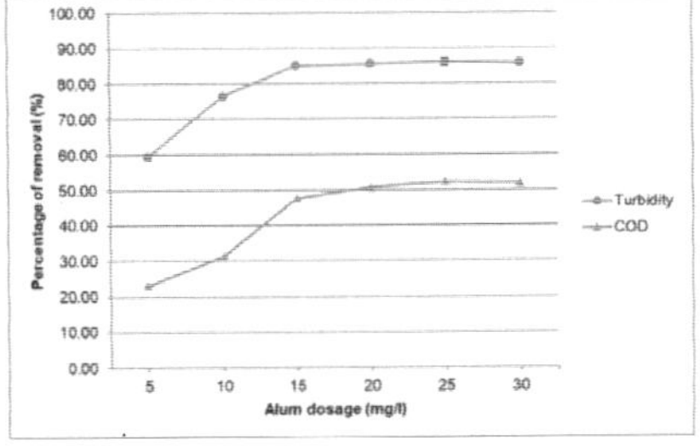

Fig. 6. Effect of Alum Dosage

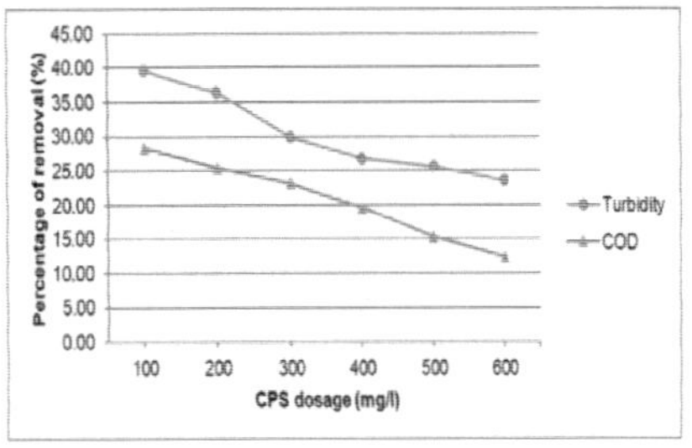

Fig. 7. Effect of CPS Dosage

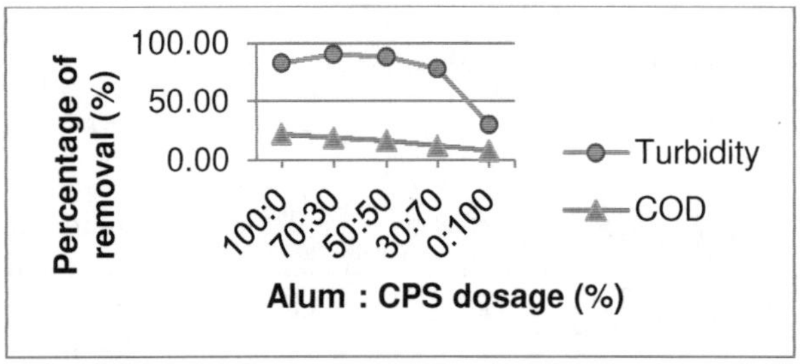

Fig. 8. Effect of Alum + CPS Dosage

4.Conclusions

This results from this study revealed that Cassava peel has the characteristic to be coagulant aid in removing turbidity and chemical oxygen demand.

5.Acknowledgement

The authors would like to acknowledge Ministry of Higher Educatin and UTHM for funding this research under FRGS 1575.

6. References

[1] J.D Theodoro, (2013). Coagulants and Natural Polymers:Perspectives for the Treatment of Water. Plastic and Polymer Technology (PAPT) Volume 2 Issue 3.

[2] N. Zayadi, N. Othman., R. Hamdan, (2016). A potential waste to be selected as media for metal and nutrient removal, IOP Conference Series : Material Science and Engineering, 136 (1) 012051

[3] S.Y. Choy (2014). Utilization of plant-based natural coagulants as future alternatives towards sustainable water clarification. Journal of Environmental Science 2 6, 2178 - 2189.

[4] Zayadi, N and Othman, N.(2013). Characterization and optimization of heavymetals biosorption by fish scale. Advanced Material Research. 795: 260-265

[5] J. Zhang, (2006). A preliminary study on cactus as coagulant in water treatment. Else, Process Biochemistry 41, 730–733.

[6] S. Zhao. (2013). Influence of a new coagulant aid-Enteromorpha extract on coagulation performance and floc characteristics of aluminum sulfate coagulant in kaolin–humic acid solution treatment, Colloids and Surfaces A: Physicochem. Eng. Aspects 417, 161– 169.

[7] Saritha, K. S. (2012). Evaluation of Chitin as Natural Coagulant in Water Treatment. Journal of Advanced Laboratory Research in Biology, 109-114.

[8] N.Othman, S.Y Kueh, M-F-H Azizul Rahman, R, Hamdan,R. Watermelon rind: a potential adsorbent for zinc removal, Applied mechanics and material, 680, 146-149 (2014)

[9] N. Othman, S. Mohd-Asharuddin, *Cucumis melo* rind as biosorbent to remove Fe(II) and Mn(II) from synthetic groundwater solution, Advanced Materials Research, **795**, 266-271, (2013).

[10] M.J. Fakir (2012). Starch and flour extraction and nutrient composition of tuber in seven. J. Bangladesh Agril. Univ. 10(2), 217-222.

IOP Conf. Series: Materials Science and Engineering **311** (2018) 012023 doi:10.1088/1757-899X/311/1/012023

CO_2 Hydrogenation over Ru/χ-Al_2O_3 Catalyst

Kanyanat Jeenjumras[1], Sunthon Piticharoenphun[1], Okorn Mekasuwandumrong[1]

1Department of Chemical Engineering, Faculty of Engineering and Industrial Technology, Silpakorn University, Nakorn Pathom 73000, Thailand
E-mail: jeenjumras_k@hotmail.com, manmsp@hotmail.com, okornm@yahoo.com

Abstract. The hydrogenation of CO_2 was investigated over Ru/χ-Al_2O_3 catalyst prepared by thermal decomposition of gibbsite which was followed with impregnation using different types of ruthenium precursors. The performance of the catalysts was examined by the temperature programmed reaction of CO_2 with H_2. The Ru/χ-Al_2O_3 catalyst prepared using ruthenium (III) nitrosyl nitrate solution as the Ru source (RNN) exhibited the best performance, compared to other Ru precursors. The physiochemical properties of each catalyst were characterized using XRD diffraction, N_2 physisorption and H_2 chemisorption.

1. Introduction

In the past centuries, the amount of carbon dioxide (CO_2) in the atmosphere has increased continuously due to human activities and the combustion of fossil fuels in the industrial processes. The increase in CO_2 concentration is a key factor in the greenhouse effect, causing the rising global average temperature and climate change. Therefore, the study of potential strategies to reduce the amount of CO_2 in atmosphere such as storage of CO_2, reduction of CO_2 produced and conversion of CO_2 for usage is urgently required [1-3].

The hydrogenation of CO_2 is an interesting option to reduce greenhouse effects as, in the presence of hydrogen, it converts CO_2 with the presence of hydrogen into hydrocarbon such as methane (CH_4) accompanied by water. CO_2 can be captured from air, human activities and industrial processes by adsorption, and then CO_2 is converted into methane. Methane is a major component of synthetic natural gas (SNG) which can be utilized to produce energy through combustion. Moreover, CO_2 derived from energy production can be re-converted into methane using CO_2 hydrogenation. The reaction of CO_2 hydrogenation for methane production is shown below (equation (1)) [4,5].

$$CO_2 + 4H_2 \leftrightarrow CH_4 + 2H_2O, \ \Delta H = -164 \text{ kJ/mol} \qquad (1)$$

As the reaction of CO_2 hydrogenation (equation (1)) is strongly exothermic, it thermodynamically favors low temperatures. Thereby, CO_2 hydrogenation reactions are highly active at low temperatures and catalysts can endure the heat of reaction. Among different types of catalyst, Ru is regarded as the most active catalyst and the most stable noble metal [6,7]. Generally, alumina is one of the most interesting catalyst supports due to its varieties of crystalline phases, which possess specific properties for particular applications. Among all Al_2O_3, χ-Al_2O_3 is one of low temperature transition alumina, which offers high surface area and thermal stability. In addition, it has high potential to be used as the catalyst and/or the support [8,9].

In this work, χ-Al_2O_3 was prepared using the thermal decomposition of microcrystalline gibbsite. The prepared χ-Al_2O_3 was used as the support for the Ru supported catalyst in the CO_2 hydrogenation reaction. The effect of Ru precursors on the physiochemical and catalytic properties of Ru/χ-Al_2O_3 was investigated.

2. Experimental

2.1Catalyst preparation

χ-Al_2O_3 support was prepared using the thermal decomposition of microcrystalline gibbsite (Sigma-Aldrich) at high temperatures. The gibbsite was calcined in a tube furnace with air flow (100 ml/min) at 600°C for 4 hours with a heating rate of 10°C/min. The 1 % Ru/χ-Al_2O_3 catalyst used in this study was synthesized using the incipient wetness impregnation techniques. 2 g of χ-Al_2O_3 was impregnated with 3 different types of ruthenium precursors, viz., ruthenium (III) acetylacetonate ($Ru(C_5H_7O_2)$, 97%, Sigma-Aldrich), ruthenium (III) nitrosyl nitrate solution ($Ru(NO)(NO_3)_x(OH)_y$, x + y = 3, 1.5% Ru in nitric acid, Sigma-Aldrich) and hexaammineruthenium (III) chloride ($Ru(NH_3)_6Cl_3$, 98%, Sigma-Aldrich), respectively. The impregnated powder was kept at room temperature for 6 hours and dried at 110°C for 12 hours. After that, the dried powder was calcined in the air at 300°C for 4 hours with a heating rate of 10°C/min. The catalysts prepared using ruthenium (III) acetylacetonate, ruthenium(III) nitrosyl nitrate solution and hexaammineruthenium (III) chloride were symbolized as RAA, RNN and RCl, respectively.

2.2Activity test

The activity test of the catalysts was performed in a continuous-flow fixed-bed reactor with the inner diameter of 7.1 mm (stainless steel) at atmospheric pressure. 0.15 ml of the catalyst powder was used in the reaction, using temperature programmed reaction method. The catalysts were reduced using H_2 with the flow rate of 30 ml/min at 400°C for 2 hours before the reaction start. The feed gas mixture of $H_2/CO_2 = 3$ with the balance helium were used in the reaction. GHSV was calculated as the volumetric flow rate versus catalyst volume. Therefore, GHSV was at 14,400 h^{-1}. The gaseous products were measured using a gas chromatograph (SHIMADZU GC-14B) with a TCD and FID detector.

2.3Catalytic characterization

The specific surface area, average pore size and total pore volume of the fresh catalyst were measured by nitrogen gas adsorption at liquid nitrogen temperature (-196°C) using a BET SORP mini II. X-ray diffraction (XRD) patterns were determined using a Siemens D5000. The diffraction patterns were performed over the 2θ range between 20° and 80° using a Cu K_α radiation with a Ni filter. H_2 chemisorption was measured using a Micrometics AutoChem 2910 instrument. In each analysis, 60 mg of fresh catalyst was reduced using H_2 (25 ml/min) at 400°C for 2 hours. After the reduction, the sample was cooled to 100°C and then analysed at this temperature. H_2 uptake was determined using the pulse injection of H_2 into the reduced catalyst.

3. Results and Discussion

The BET surface area, pore volume and average pore diameter of the Ru/χ-Al_2O_3 catalysts prepared using each ruthenium precursors are summarized in table 1. Among the three catalysts, RAA exhibited the highest BET surface area and smallest average pore diameter, while RCl possessed the lowest BET surface area and largest average pore diameter. From XRD result, it clearly appears the peak intensity of RuO_2 in the three catalysts. And all catalysts exhibited the typical pattern of χ-Al_2O_3.

The relative amounts of active surface Ru metal on the catalyst samples was calculated from H_2 chemisorption based on the assumption that one hydrogen molecule adsorbs on two ruthenium sites. The results of H_2 chemisorption over various ruthenium precursor catalysts are presented in table 1. The dispersion and average particle size of Ru was calculated using the following equation (2).

$$\%\text{Dispersion} = (V_{STP}/22414) \times S \times MW \times 100/(\%\text{ metal weight}) \qquad (2)$$

Where H_2 coveraged on Ru at STP is represented as VSTP and stoichiometric factor of H2 to atom of Ru is assigned as S = 2. The Ru metal dispersion was improved in order of RNN > RCl >RAA.

Table 1. BET surface area, pore volume, average pore diameter and H_2 chemisorption of Ru/χ-Al_2O_3 catalysts prepared using each ruthenium precursor.

Catalyst	Surface area [$m^2\ g^{-1}$]	Pore volume [$cm^3\ g^{-1}$]	Average pore diameter [nm]	H_2 uptake [$\mu mol\ g^{-1}$]	Dispersion [%]
RCl	155	0.21	5.5	23.8	24.10
RNN	171	0.21	5.0	31.0	31.31
RAA	192	0.19	4.0	5.6	5.62

The morphology of Ru/χ-Al_2O_3 catalysts are determined using TEM method and the results are shown in figure 1. It was found that the different RuO_2 particles were dispersed on the χ-Al_2O_3 surface. RNN and RCl catalyst exhibited the polyhedral shape particles of RuO_2, whereas, RAA catalyst appeared rod-shape particles.

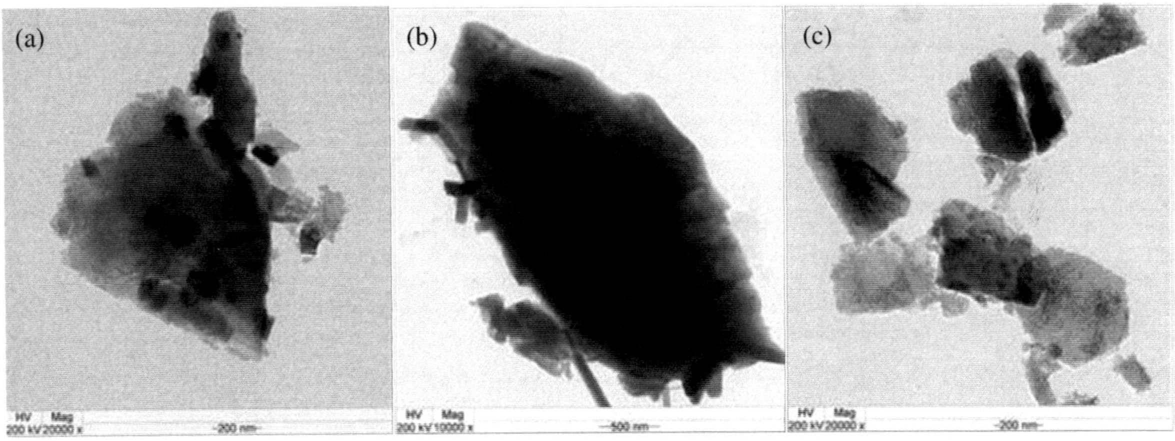

Figure 1. TEM images of Ru/χ-Al_2O_3 catalysts using different ruthenium precursors: (a) RNN, (b) RAA, (c) RCl.

The performance of the Ru/χ-Al_2O_3 catalysts was investigated using CO_2 hydrogenation at atmospheric pressures and different reaction temperatures between 250 and 425°C (with increasing steps of 25°C). The graph plotted between catalytic performance and reaction temperature for Ru/χ-Al_2O_3 catalysts with different ruthenium precursors is shown in figure 2. It can be seen that the increase in CO_2 conversion depended on the increasing reaction temperature. The RNN catalyst obtained the highest CO_2 conversion profile, compared to the other two catalysts. CO_2 conversion of the RNN catalyst started at 250°C and monotonically increased and then slowed down from 375°C onwards. For the RAA catalyst, the CO_2 conversion profile was slightly lower than the RNN catalyst. The RCl catalyst showed the lowest CO_2 conversion profile and its CO_2 conversion started at 275°C and slowly increased with the increasing reaction temperature.

The selectivity of the CH_4 profiles of the RCl catalyst was much lower than the RNN and RAA catalysts. In addition, the CH_4 selectivity of the RNN catalyst was indistinguishable from the RAA catalyst. The CH_4 yield showed that the formation rate of methane rose with the increasing reaction temperature. The RNN catalyst gave maximum CH_4 yield which was higher than the other catalysts. Moreover, CO formation started at the same temperature (~300°C) for all Ru precursor catalysts.

As mentioned above, the conversion of CO_2 into methane favored low temperatures and the maximum CO_2 conversion was consistent with the maximum methane formation rate. At high temperatures, the result indicated that there was the appearance of CO formation. This behaviour occurs because of the reverse water gas shift (equation (3)).

$$CO_2 + H_2 \rightarrow CO + H_2O \quad (3)$$

From the catalytic performance results, RNN exhibited the best performance due to its highest Ru dispersion and also smallest Ru particle size, calculated from H_2 chemisorption. However, RCl also possessed high Ru dispersion but exhibited the lowest performance. This was probably due to the remains of residual chlorine, retarding the adsorption of feed gas on the catalyst surface [10].

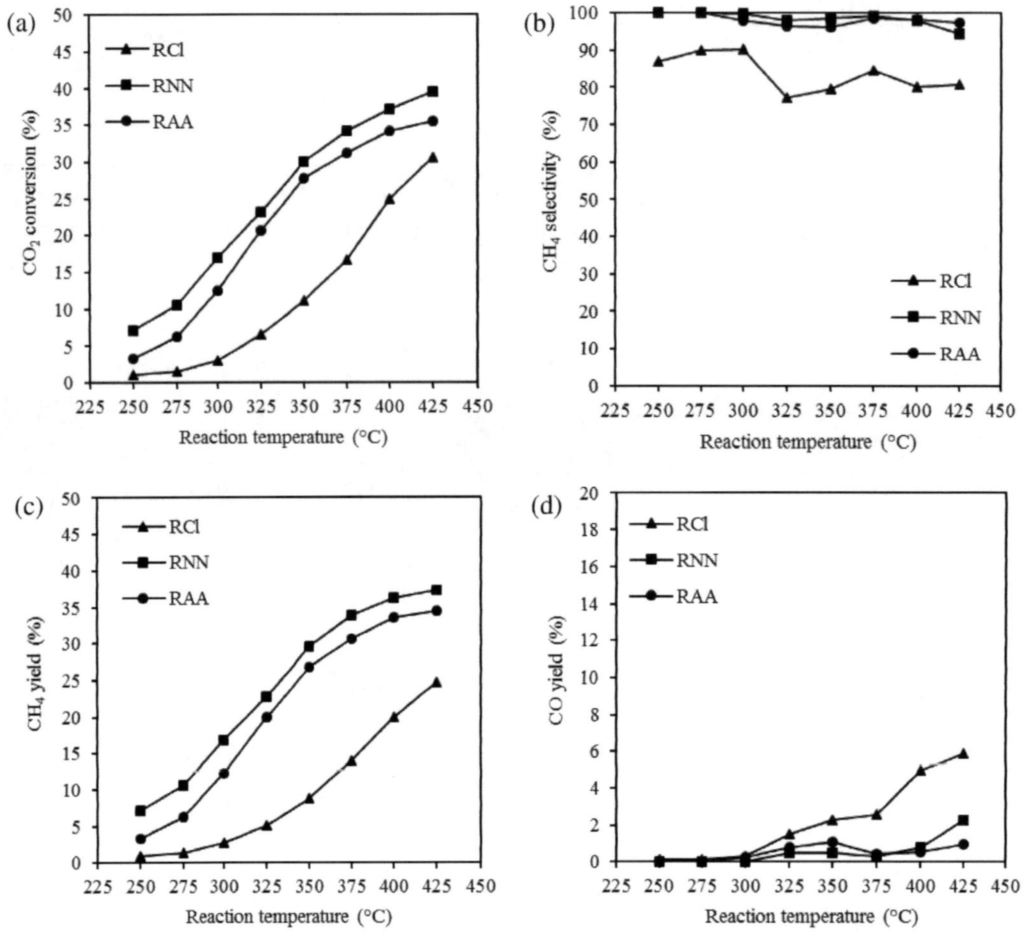

Figure 2. The performance of Ru/χ-Al_2O_3 catalysts as a function of the reaction temperature using different ruthenium precursors: (a) CO_2 conversion, (b) CH_4 selectivity, (c) CH_4 and (d) CO yields.

4. Conclusion

The effect of ruthenium precursors in Ru/χ-Al_2O_3 on the catalytic performance in CO_2 hydrogenation reaction was performed using a temperature programmed reaction. The CO_2 conversion of the RNN catalyst was about 2 times higher than that of the RCl catalyst with a very small amount of byproduct at 375°C. The RCl catalyst showed low activity, probably due to the remains of residual chlorine. As the dispersion was highly effective, the Ru particle of the RNN catalyst obtained the smallest size among all the catalysts. Moreover, it gave higher CO_2 conversion and CH_4 formation than the other two ruthenium precursor catalysts. Although the RNN catalyst does not significantly differ from the RAA catalyst, it can be safely assumed that the catalytic performance is greater. As mentioned above, it is clear that the types of ruthenium precursors affect the activity and physical properties of catalysts.

5. Acknowledgments

Financial supports from Silpakorn University Research and Development Institute are gratefully acknowledged.

IOP Conf. Series: Materials Science and Engineering **311** (2018) 012023 doi:10.1088/1757-899X/311/1/012023

6. References

[1] W Wang, S P Wang, X B Ma and J L Gong 2011 *Chem. Soc. Rev.* **40** 3703–3727
[2] S Solomon, G -K Plattner, R Knutti and P Friedlingstein 2009 *Proc. Natl. Acad. Sci.* **106** 1704–1709
[3] J C M Pires, F G Martins, M C M Alvim-Ferraz and M Simoes 2011 *Chem. Eng. Res. Des.* **89** 1446–1460
[4] C Janke, M S Duyar, M Hoskins and R Farrauto 2014 *Appl. Catal. B Environ.* **152–153** 184–191
[5] Suhas G Jadhav, Prakash D Vaidya, Bhalchandra M Bhanage and Jyeshtharaj B Joshi 2014 *Chem. Eng.* Res. *Des.* **92** 2557–2567
[6] J H Xu, Q Q Lin, X Su, H M Duan, H R Geng and Y J Huang 2016 *Chin. J. Chem. Eng.* **24** 140–145
[7] D Li, N Ichikuni, S Shimazu and T Uematsu 1998 *Appl. Catal. A Gen.* **172** 351–358
[8] W Chaitree, S Jiemsirilers, O Mekasuwandumrong, B Jongsomjit, A Shotipruk and J Panpranot 2011 *Catal. Today* **164** 302–307
[9] O Mekasuwandumrong, P Silveston, P Praserthdam, M Inoue, V Pavarajarn and W Tanakulrungsank 2003 *Inorg. Chem. Commun.* **6** 930–934
[10] S Murata and K Aika 1992 *J. Catal.* **136** 110–117

IOP Conf. Series: Materials Science and Engineering **311** (2018) 012024 doi:10.1088/1757-899X/311/1/012024

Effect of tween 80 on nanoparticle preparation of modified chitosan for targeted delivery of combination doxorubicin and curcumin analogue

Anita Sukmawati[1], Wahyu Utami[1], Ratna Yuliani[1], Muhammad Da'i[1] and Akhmad Nafarin[1]

[1]Faculty of Pharmacy, Universitas Muhammadiyah Surakarta, Pabelan, Surakarta, 57162, Indonesia

anita.sukmawati@ums.ac.id

Abstract. Delivery of anticancer is facing several problems including unspecific delivery of active substance to the targeted cell. The conjugation between chitosan and folate (chitosan-FA) was used for nanoparticle preparation containing combination of doxorubicin (DOX) and curcumin analogue, *2,5-bis-(4-hydroxi,3,5-dimethyl)-benzylidincylopentanone,* as active substances. The purpose of this research is investigating formulation aspect for chitosan-FA nanoparticle by addition various tween 80 to achieve desired nano-size particle. The ionic gelation method was used for nanoparticle preparation using 0.05% *w/v* chitosan-FA with addition of 0.1 and 0.5% *v/v* of tween 80. The result showed that the high concentration of tween 80 during nanoparticle preparation lead to formation of smaller size particle. The 111.8 ±4.11 nm particle size was revealed by addition of 0.5% *v/v* tween 80 during chitosan-FA nanoparticle preparation loaded with active substances.

1. Introduction

Delivery of active agent for cancer therapy has several challenging aspect including how to specifically bring the agents to the targeted cell, particularly when the cancer cell spread to the body, an anticancer therapy is needed for the patient [1]. However, the common problem related to anticancer therapy is several drawbacks including gastrointestinal and immune system problem causing by unspecific delivery of anticancer agents. Thus, cancer therapy using specific targeted therapy should be developed. Nanoparticle can be used for delivering an anticancer since the ability of the system for protecting the active substance and the nano-size particle would improving the penetration across the mucosal epithelium [2,3]. Moreover, the nano-size particle usually provide a larger surface area for attachment with the cell and shorter path for drug diffusion to the target cell [4].

One of potential polymer for drug delivery is chitosan, a deacetylated chitin based material. This material offers several advantages including low toxicity, biodegradable and has immuno-stimulating effect [2]. Moreover, the amine primer groups on chitosan can be chemically modified by attaching with specific ligand for targeted delivery. Drug incorporation into nano-polymeric system also can be used for controlled release of drug over the extended period and functionalized with appropriate ligand for targeted delivery of active substance [5].

In order to minimize the toxicity of drug, combination of potent anticancer can be used since it will reduce the required dose of each drug. In this research, doxorubicin (DOX), a potent anticancer, is combined with a curcumin analogue, *2,5-bis-(4-hydroxi,3,5-dimethyl)-benzylidincylopentanone* or pentagamavuno-1 (PGV-1) for inducing apoptosis in cancer cell as described in previous research by Hermawan and Meiyanto [6,7]. Nano particulate system using chitosan is proposed for delivering these agents since the chitosan nanoparticle (NP) can be selectively delivered anticancer agents to the tumour cell and has enhance permeation and retention (EPR) effect inside the cell [1,8]. However, retention and permeation of the particle inside the cell strongly influenced by tumour type, location and vascular density of tumour and further affect the effectivity to the cancer cell [10]. As consequences, the modification of chitosan nanoparticle is needed to achieve specific delivery to cancer cell.

Folic acid (FA) is proposed to use as a ligand for specific delivery of nanoparticle since it enhancing endocytosis of particle through folate receptor [11]. Additionally, FA is a relative stable ligand, has lower immunogenic activity and has high affinity with folate receptor. Nanoparticle conjugate with folic acid entered the cancer cell via folate receptor-mediated endocytosis and transferred through cell's organelles by vesicular trafficking and then releases the active substances into cell cytoplasm [11]. The conjugation of FA into chitosan polymer can be performed through selective amidation of primary amine of chitosan as described by Yang *et al* [11].

The physical characteristics of particle including the particle size, is strongly influenced by nanoparticle preparation [13,14]. Further consequences, it will affect the activity of the nanoparticle inside the cancer cell. This research will explore the preparation and characterisation of chitosan modified folic acid nanoparticle. The conjugation of FA into chitosan polymer would be characterised using spectrophotometer UV. Modified chitosan nanoparticle was prepared using ionic gelation method using 0.1% of sodium tripolyphosphate (Na TPP). Tween 80 as surfactant was added during the nanoparticle preparation in order to achieve the desired size of particle. The effect of various concentration of tween 80 towards particle size was investigated.

2. Material and Methods

2.1 Material.

Chitosan (95% deacetylation, 150 kDa, pharmaceutical grade) purchased from CV. ChiMultiguna, Indonesia, *2,5-bis-(4-hydroxy, 3,5-dimethyl)-benzylidine-cyclopentanone* (pentagamavunon-1, PGV-1) purchased from Faculty of Pharmacy, Universitas Gadjah Mada, Indonesia, sodium tripolyphosphate (Na TPP), sodium hydroxide (NaOH) and acetic acid were purchased from Bratachem, Indonesia. Folic acid (FA), *N-(e-dimethylaminopropyl)-N-ethylcarbodiimide hydrochloride* (EDC hydrochloride), dimethyl sulfoxide (DMSO), doxorubicin hydrochloride (DOX-HCl) and phosphate buffer saline (PBS) tablet were purchased from Sigma-Aldrich.

2.2 Preparation of chitosan modified folic acid.

Conjugation of folic acid (FA) to chitosan was prepared using the method as described in Yang, *et al* (2010) [11]. The FA was attached to chitosan with 0.2 mol ratio to 1 mol chitosan. A 250 mg of chitosan (1.55 mmol) was dissolved in 50 ml of 1M acetic acid. The 20 mL DMSO solution containing 136.83 mg (0.31 mmol) of FA and 59.43 mg (0.31 mmol) of EDC hydrochloride then added to chitosan solution and continuously stirred using magnetic stirrer for 18 hour in the dark. After 18 hour, 1 M sodium hydroxide was dripped into the solution to bring the pH to 9. The mixture then centrifuged at 2500 rpm to precipitate the chitosan modified folic acid. Chitosan modified folic acid (chitosan-FA) then dissolved in 50 ml water and dialyzed against phosphate buffer saline (PBS) for 3 days then continuing dialyzed against water for 4 days. The solution containing chitosan-FA then freeze dried to get the sponge mass of chitosan-FA and stored at 4°C.

2.3 Confirmation of chitosan modified folic acid.

A 10 mg of chitosan-FA was dissolved in 10 ml of 1M acetic acid. The solution then diluted to 100 μg/ml using water and scanned with spectrophotometer UV-VIS (Genesys 10S) at 200-300 nm. The measured absorbance spectrum was compared with chitosan solution in acetic acid, folic acid solution in water and EDC hydrochloride in water.

2.4 Preparation of chitosan-FA nanoparticle.

The nanoparticle of chitosan-FA was prepared using a method as described in Zhang *et al* [15] with modification. A 100 mg of chitosan-FA was dissolved in 200 ml of 0.5% acetic acid solution containing 0.1 and 0.5% *v/v* of tween 80 to give polymer concentration 0.05% w/v. The mixture was stirred for 3 minutes at 1500 rpm using magnetic stirrer (IKA). Afterwards, the pH was measured and adjusted to 4 using 1N sodium hydroxide. A 20 ml of 0.1% Na TPP then added by slow dripping under magnetic stirrer to form nanoparticle. The mixture then continued to stir for 4 hours and isolated by centrifugation at 12,500 rpm. The precipitated mass then freeze dried to give dry particle and keep in the dark at 4°C.

The chitosan-FA nanoparticle loading with drugs was prepared using the method described above with addition of 1 ml DOX (2 mg/mL) and PGV-1(10 mg/ml) to give the final concentration of 2% *w/w* and 10%

w/w drugs to polymer, respectively. Chitosan nanoparticle also prepared using the same method in preliminary experiment. The variables applied for the nanoparticle formulation is given in **Table 1.**

Table 1. Formulation of chitosan and chitosan-FA nanoparticles

Polymer	Tween 80 (%*v/v*)	Active substances
Chitosan	-	DOX-PGV-1
Chitosan-FA	-	-
Chitosan-FA	0.1	-
Chitosan-FA	0.5	-
Chitosan-FA	0.5	DOX-PGV-1

Evaluation of particle size. Particle size and polydispersity index of nanoparticles were evaluated by laser diffraction analysis using particle size analysis (PSA, Horiba SZ100) at 25±0.5°C. Ten milligram of nanoparticle was dispersed in aqua pro injection and directly measured using PSA to obtained particle size and polydispersity index. All measurements were performed triplicate.

3. Result and Discussion

3.1 Preparation and identification of chitosan modified folic acid.

The chitosan modified folic acid (chitosan-FA) was successfully synthesized with 53.2±0.02 % of yield. Sequential process involved during chitosan-FA preparation including purification by dialysis might reduce the yield of chitosan-FA. During dialysis process, the unbound folic acid (FA) and soluble EDC hydrochloride diffused outside of the dialysis membrane and removed from chitosan-FA.

Conjugation of chitosan FA as well as purification process was confirmed via UV absorption spectra. The UV absorption spectra of chitosan, FA, EDC hydrochloride are shown in **Fig. 1**.

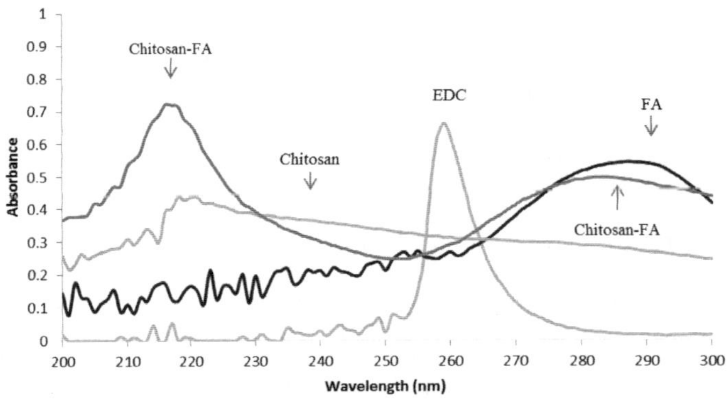

Fig. 1. Confirmation of chitosan-FA conjugation and purification by UV absorption spectroscopy. Chitosan (2.5 mg/ml) and chitosan-FA (100 µg/ml) were measured in 1 M acetic acid solution while folic acid (2 µg/ml) and EDC hydrochloride (10 µg/ml) were measured in water.

Chitosan has maximum absorbance at λ 218 nm, related to chromophore groups from N-acetyl glucosamine and glucosamine [16]. However, the absorption peak was weak compared to chitosan-FA. Chitosan-FA absorbance reached maximum at λ 216 nm and 282 nm with high intensity related to n →π* and π → π* transition corresponding to amide bond formation and C=C bond transition in chitosan-FA [17]. FA itself has the maximum absorbance at λ 287 nm related to π → π* of aromatic ring in folic acid. The unpresented of EDC hydrochloride peak in chitosan-FA revealed that the dialysis method can be used for chitosan-FA purification.

3.2 Preparation and particle size determination of nanoparticles.

Ionic gelation method was used for nanoparticle preparation by interaction of positive charge of chitosan and negative charge of Na TPP. In this research, the effect of tween 80 on particle size was investigated. For nanoparticle delivery purposes, the size of particle may vary from 10 nm-1000 nm. However, the smaller size of particle will be easier to uptake by the cell and has a larger surface area. As consequences, a smaller particle will be related to faster drug release [18].

In order to obtain desired nano-size particle, at the beginning of experiment, the chitosan-FA nanoparticle was prepared without loaded with active substances. It can be seen at Table 2. that the chitosan-FA NP without tween 80 lead to the formation of micro-size particle and high polydispersity index (PI). However, preliminary studies using the same method showed that the method can be used to produce nano-size particle with polydispersity index under 0.7 indicating homogeneity of particle. This diverse result might relate to folate conjugation into chitosan molecule. Addition of folate molecule into chitosan backbone affected chitosan molecular properties. Chitosan-FA had higher molecular size than chitosan itself. High molecular size of polymer reveled to produce bigger particle size as evaluated in earlier experiment by Caetano *et al* and Sarwar *et al* [19, 20].

Table 2. Particle size and polydispersity index of chitosan and chitosan-FA nanoparticles

Polymer	**Tween 80 [*%v/v*]**	**Active substances**	**Particle size [nm] ± SD**	**Polydispersity index [PI] ± SD**
Chitosan	-	DOX-PGV-1	812.8 ± 11.94	0.50 ± 0.21
Chitosan-FA	-	-	5581.4 ± 33.67	4.25 ± 1.92
Chitosan-FA	0.1	-	554.2 ± 40.46	0.44 ± 0.12
Chitosan-FA	0.5	-	476.6 ± 11.94	0.49 ± 0.06
Chitosan-FA	0.5	DOX-PGV-1	111.8 ±4.11	0.50 ± 0.21

Tween 80 addition during nanoparticle preparation improved the particle size of chitosan FA into nano-scale (Table 2). Nevertheless, the addition of 0.5% v/v of tween 80 leading to the formation of smaller particle size compared to 0.1% v/v tween 80. The result indicated that tween 80 might act as stabilizing agent during nanoparticle formation and reducing the surface energy leading to inhibition of crystal growth [21]. Conversely, the experiment by Asasutjarit *et al* reported that increased tween 80 concentration ranged from 0 to 0.03% *v/v* acted as amphiphilic molecule and can be deposited at the surface of particle resulting in rising of particle size [22]. It can be determined that different range of tween concentration applied during nanoparticle preparation could be altered the size of particle by diverse effect.

The 0.5% v/v tween 80 then continue to use for preparing chitosan-FA loaded with active substances, DOX and PGV-1. Evaluation on particle size showed that the size of particle was 111.8 ±4.11 nm with the addition of active substances. This size was lower than unloaded chitosan-FA prepared using the same method. Generally, the size of particle will be bigger with the presence of active substance although the increasing the particle size are not significant [23, 24]. This unexpected result might be related to the addition of active substances, mainly DOX, since DOX added to the formula in was the form of doxorubicin hydrochloride. The presence of hydrochloride salt from DOX in nanoparticle preparation might contribute to solubility of the chitosan-FA in acetic acid, hence, the fine particles were formed. However, the other aspect related to nanoparticle morphology, encapsulation efficiency of active substances and drug release should be investigated.

4. Conclusion

This study present that chitosan-FA was successfully conjugated and purified as confirmed by UV absorbance. Moreover, this research present that addition tween 80 affected the size of chitosan-FA nanoparticle. The higher concentration tween 80 applied during formulation decreased the size of nanoparticle. The desired nanosize particle was achieved using 0.5% v/v tween 80 and need further identification using scanning electron microscope.

5. Acknowledgment

We would gratefully thanks to Ministry of Research, Technology and Higher Education of Indonesia (RISTEKDIKTI) for funding of this research. We also thanks to Universitas Muhammadiyah Surakarta, Indonesia for giving facility to do this research and Faculty of Pharmacy, Universitas Gadjah Mada, Indonesia for kindly help to provide PGV-1 for this research.

6. References

[1] S. Bisht and A. Maitra, "Nanoparticles for Solid Tumor," *Cancer*, vol. 1, no. 4, pp. 415–425, (2009).

[2] A. Rampino, M. Borgogna, P. Blasi, B. Bellich, and A. Cesàro, "Chitosan nanoparticles: Preparation, size evolution and stability," *Int. J. Pharm.*, vol. 455, no. 1–2, pp. 219–228, (2013).

[3] K. A. Janes, M. P. Fresneau, A. Marazuela, A. Fabra, and M. J. Alonso, "Chitosan nanoparticles as delivery systems for doxorubicin," *J. Control. Release*, vol. 73, no. 2–3, pp. 255–267, Jun. (2001).
[4] L. Zhao, L. Shi, Z. Zhang, J. Chen, D. Shi, J. Yang, and Z. Tang, "Preparation and Application of Chitosan Nanoparticles and Nanofibers," vol. 28, no. 3, pp. 353–362, (2011).
[5] M. Conti, V. Tazzari, C. Baccini, G. Pertici, L. P. Serino, and U. De Giorgi, "Anticancer drug delivery with nanoparticles.," *In Vivo*, vol. 20, no. 6A, pp. 697–701, (2006).
[6] A. Hermawan, A. Fitriasari, S. Junedi, M. Ikawati, S. Haryanti, B. Widaryanti, M. Da'i, and E. Meiyanto, "PGV-0 and PGV-1 Increased Apoptosis Induction of Doxorubicinon MCF-7 Breast Cancer Cells," *Pharmacon*, vol. 12, no. 2, pp. 55–59, (2011).
[7] E. Meiyanto, D. D. P. Putri, R. A. Susidarti, R. Murwanti, S. Sardjiman, A. Fitriasari, U. Husnaa, H. Purnomo, and M. Kawaichi, "Curcumin and its Analogues (PGV-0 and PGV-1) Enhance Sensitivity of Resistant MCF-7 Cells to Doxorubicin through Inhibition of HER2 and NF-kB Activation," *Asian Pacific J. Cancer Prev.*, vol. 15, no. 1, pp. 179–184, (2014).
[8] Y. Xu, Z. Wen, and Z. Xu, "Chitosan Nanoparticles Inhibit the Growth of Human Hepatocellular Carcinoma Xenografts through an Antiangiogenic Mechanism," *Anticancer Res.*, vol. 30, pp. 5103–5110, (2010).
[9] Y. S. Wimardhani, D. F. Suniarti, H. J. Freisleben, S. I. Wanandi, N. C. Siregar, and M. Ikeda, "Chitosan exerts anticancer activity through induction of apoptosis and cell cycle arrest in oral cancer cells," vol. 56, no. 2, pp. 119–126, (2014).
[10] V. M. Gaspar, E. C. Costa, J. A. Queiroz, C. Pichon, F. Sousa, and I. J. Correia, "Folate-targeted multifunctional amino acid-chitosan nanoparticles for improved cancer therapy," *Pharm. Res.*, vol. 32, no. 2, pp. 562–577, (2014).
[11] S. Yang, F. Lin, K. Tsai, M. Wei, H. Tsai, and J. Wong, "Folic Acid-Conjugated Chitosan Nanoparticles Enhanced Protoporphyrin IX Accumulation in Colorectal Cancer Cells - Bioconjugate Chemistry (ACS Publications)," pp. 679–689, 2010.
[12] C. Zhang, S. Gao, W. Jiang, S. Lin, F. Du, Z. Li, and W. Huang, "Targeted minicircle DNA delivery using folate-poly(ethylene glycol)-polyethylenimine as non-viral carrier," *Biomaterials*, vol. 31, no. 23, pp. 6075–6086, (2010).
[13] P. Calvo, J. L. Vila-Jato, and M. J. Alonso, "Effect of lysozyme on the stability of polyester nanocapsules and nanoparticles: stabilization approaches," *Biomaterials*, vol. 18, no. 19, pp. 1305–1310, (1997).
[14] A. Nasti, N. M. Zaki, P. De Leonardis, S. Ungphaiboon, P. Sansongsak, M. G. Rimoli, and N. Tirelli, "Chitosan/TPP and chitosan/TPP-hyaluronic acid nanoparticles: Systematic optimisation of the preparatıve process and preliminary biological evaluation," *Pharm. Res.*, vol. 26, no. 8, pp. 1918–1930, (2009).
[15] H. Zhang, M. Oh, C. Allen, and E. Kumacheva, "Monodisperse Chitosan Nanoparticles for Mucosal Drug Delivery Monodisperse Chitosan Nanoparticles for Mucosal Drug Delivery," *Biomacromol*, vol. 5, pp. 2461–2468, (2004).
[16] J. Kumirska, Z. Kaczy, and A. Bychowska, "Application of Spectroscopic Methods for Structural Analysis of Chitin and Chitosan," pp. 1567–1636, (2010).
[17] N. L. Dhas, P. P. Ige, and R. R. Kudarha, "Design, Optimization and in-vitro study of Folic acid Conjugated-Chitosan Functionalized PLGA Nanoparticle for Delivery of Bicalutamide in Prostate Cancer," *Powder Technol.*, vol. 283, pp. 234–245, (2015).
[18] R. Singh, J. W. Lillard, and Jr., "Nanoparticle-based targeted drug delivery.," *Exp. Mol. Pathol.*, vol. 86, no. 3, pp. 215–23, Jun. (2009).
[19] L. A. Caetano, A. J. Almeida, and L. M. D. Gon??alves, "Effect of experimental parameters on alginate/chitosan microparticles for BCG encapsulation," *Mar. Drugs*, vol. 14, no. 5, (2016).
[20] A. Sarwar, H. Katas, and N. M. Zin, "Antibacterial effects of chitosan-tripolyphosphate nanoparticles: Impact of particle size molecular weight," *J. Nanoparticle Res.*, vol. 16, no. 7, (2014).
[21] S. Rajaram and R. Natham, "Influence of formulation and process variables on the formation of rifampicin nanoparticles by ionic gelation technique," *Res. J. Pharm. Biol. Chem. Sci.*, vol. 4, no. 4, pp. 820–832, (2013).
[22] R. Asasutjarit, C. Sorrachaitawatwong, N. Tipchuwong, and S. Pouthai, "Effect of formulation compositions on particle size and zeta potential of diclofenac sodium-loaded chitosan nanoparticles," *Int. J. Medical, Heal. Biomed. Bioeng. Pharm. Eng.*, vol. 7, no. 9, pp. 309–311, (2013).
[23] H. Katas, Z. Hussain, and S. A. Awang, "Bovine Serum Albumin-Loaded Chitosan / Dextran Nanoparticles : Preparation and Evaluation of Ex Vivo Colloidal Stability in Serum," *J. Nanomater.*, vol. 2013, no. Article ID 536291, p. 9 pages, (2013).

[24] M. D. N, R. Eskandari, H. Zolfagharian, and M. Mohammad, "Preparation and in vitro characterization of chitosan nanoparticles containing Mesobuthus eupeus scorpion venom as an antigen delivery system," *J. Venom. Anim. Toxins Incl. Trop. Dis.*, vol. 18, no. 1, pp. 44–52, (2012).